Fireworks

Quadratic Functions, Graphs, and Equations

Teacher's Guide

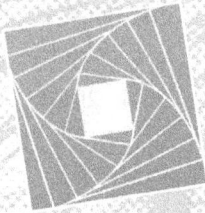

This material is based upon work supported by the National Science Foundation under award numbers ESI-9255262, ESI-0137805, and ESI-0627821. Any opinions, findings, and conclusions or recommendations expressed in this publication are those of the authors and do not necessarily reflect the views of the National Science Foundation.

Key Curriculum
1150 65th Street
Emeryville, California 94608
email: editorial@keypress.com
www.keycurriculum.com

First Edition Authors

Dan Fendel, Diane Resek, Lynne Alper, and Sherry Fraser

Contributors to the Second Edition

Sherry Fraser, Jean Klanica, Brian Lawler, Eric Robinson, Lew Romagnano, Rick Marks, Dan Brutlag, Alan Olds, Mike Bryant, Jeri P. Philbrick, Lori Green, Matt Bremer, Margaret DeArmond

Project Editor

Sharon Taylor

Consulting Editor

Mali Apple

Project Administrator

Juliana Tringali

Professional Reviewer

Rick Marks, Sonoma State University

Calculator Materials Editor

Josephine Noah

Math Checker

Carrie Gongaware

Production Editor

Andrew Jones

Production Director

Christine Osborne

Executive Editor

Josephine Noah

Mathematics Product Manager

Timothy Pope

Publisher

Steven Rasmussen

Contents

Calculator Guide and Calculator Notes

Introduction

Fireworks Unit Overview

Intent

This unit uses a variety of contexts—projectile motion, areas and volumes, the Pythagorean theorem, and economics—to develop students' understanding of quadratics functions and their representations, as well as methods for solving quadratic equations.

The central problem involves a rocket used to launch a fireworks display. The height of the rocket is described by a quadratic function, and the questions involve vertices and x-intercepts, which are fundamental features of the graphs of quadratic functions.

Over the course of the unit, students strengthen their abilities to work with algebraic symbols and to relate algebraic representations to problem situations. Specifically, they see that rewriting quadratic expressions in special ways, either in factored form or in vertex form, provides insight into the graphs of the corresponding functions. Establishing this connection between algebra and geometry is a primary goal of the unit.

Mathematics

Fireworks focuses on the use of quadratic functions to represent a variety of real-world situations and on the development of algebraic skills for working with those functions. Experiences with graphs play an important role in understanding the behavior of quadratic functions.

The main concepts and skills students will encounter and practice during the unit are summarized here.

Mathematical Modeling
- Expressing real-world situations in terms of functions and equations
- Applying mathematical tools to models of real-world problems
- Interpreting mathematical results in terms of real-world situations

Graphs of Quadratic Functions
- Understanding the roles of the vertex and x-intercept in the graphs of quadratic functions
- Recognizing the significance of the sign of the x^2 term in determining the orientation of the graph of a quadratic function
- Using graphs to understand and solve problems involving quadratic functions

Working with Algebraic Expressions
- Using an area model to understand multiplication of binomials, factoring of quadratic expressions, and completing the square of quadratic expressions

- Transforming quadratic expressions into vertex form
- Simplifying expressions involving parentheses
- Identifying certain quadratic expressions as perfect squares

Solving Quadratic Equations
- Interpreting quadratic equations in terms of graphs and vice versa
- Estimating x-intercepts using a graph
- Finding roots of an equation using the vertex form of the corresponding function
- Using the zero product rule of multiplication to solve equations by factoring

Progression

The unit begins with a graphical treatment of quadratic functions. The area model for multiplication is then used as the basis for the development of a set of manipulative skills. Students use these skills and graphs to solve problems drawn from a variety of real-world contexts. In each case, the interplay between symbolic and graphical representations of quadratic functions is emphasized. In addition, there are two POWs in the unit and the option of including a third from the list of supplemental activities.

A Quadratic Rocket

The Form of It All

Putting Quadratics to Use

Back to Bayside High

Intercepts and Factoring

Supplemental Activities

Unit Assessments

Pacing Guides

50-Minute Pacing Guide (20 days)

Day	Activity	In-Class Time Estimate
	A Quadratic Rocket	
1	*Victory Celebration*	35
	POW 10: Growth of Rat Populations	10
	Homework: *A Corral Variation*	5
2	Discussion: *A Corral Variation*	15
	Parabolas and Equations I	20
	Parabolas and Equations II	15
	Homework: *Rats in June*	0
3	Discussion: *Rats in June*	10
	Parabolas and Equations III	15
	Vertex Form for Parabolas	25
	Homework: *Using Vertex Form*	0
4	Discussion: *Using Vertex Form*	15
	Crossing the Axis	25
	Homework: *Is It a Homer?*	10
	The Form of It All	
5	Discussion: *Is It a Homer?*	15
	A Lot of Changing Sides	30
	Homework: *Distributing the Area I*	5

6	Discussion: *Distributing the Area I*	15
	Views of the Distributive Property	35
	Homework: *Distributing the Area II*	0
7	Discussion: *Distributing the Area II*	15
	Square It!	25
	Homework: *Squares and Expansions*	10
8	Discussion: *Squares and Expansions*	15
	Presentations: *POW 10: Growth of Rat Populations*	20
	POW 11: Twin Primes	10
	Homework: *Vertex Form to Standard Form*	5
9	Discussion: *Vertex Form to Standard Form*	40
	Homework: *How Much Can They Drink?*	10
10	Discussion: *How Much Can They Drink?*	15
	Putting Quadratics To Use	
	Revisiting Leslie's Flowers	30
	Homework: *Emergency At Sea*	5
11	Discussion: *Emergency At Sea*	15
	Here Comes Vertex Form	30
	Homework: *Finding Vertices and Intercepts*	5
12	Discussion: *Finding Vertices and Intercepts*	15
	Another Rocket	30
	Homework: *Profiting from Widgets*	5
13	Discussion: *Profiting from Widgets*	15

	Pens and Corrals in Vertex Form	35
	Homework: *Vertex Form Continued*	0
14	Discussion: *Vertex Form Continued*	15
	Back to Bayside High	
	Fireworks in the Sky	30
	Homework: *Coming Down* and *A Fireworks Summary*	5
15	Discussion: *Coming Down* and *A Fireworks Summary*	25
	Intercepts and Factoring	
	Factoring	25
	Homework: *Let's Factor!*	0
16	Discussion: *Let's Factor!*	15
	Solve That Quadratic!	35
	Homework: *Quadratic Choices*	0
17	Discussion: *Quadratic Choices*	25
	Presentations: *POW 11: Twin Primes*	20
	Homework: *A Quadratic Summary*	5
18	Discussion: *A Quadratic Summary*	15
	"Fireworks" Portfolio	35
19	*In-Class Assessment*	45
	Homework: *Take-Home Assessment*	5
20	Exam Discussion	30
	Unit Reflection	20

90-minute Pacing Guide (13 days)

Day	Activity	In-Class Time Estimate
1	A Quadratic Rocket	
	Victory Celebration	35
	A Corral Variation	45
	POW 10: Growth of Rat Populations	10
	Homework: *Rats in June*	0
2	Discussion: *Rats in June*	10
	Parabolas and Equations I	20
	Parabolas and Equations II	15
	Parabolas and Equations III	15
	Vertex Form for Parabolas	25
	Homework: *Using Vertex Form*	5
3	Discussion: *Using Vertex Form*	20
	Crossing the Axis	25
	Is It a Homer?	45
	The Form of It All	
4	*A Lot of Changing Sides*	30
	Distributing the Area I	35
	Views of the Distributive Property	25
5	*Views of the Distributive Property* (continued)	15
	Distributing the Area II	35
	Square It!	30
	Homework: *Squares and Expansions*	10
6	Discussion: *Squares and Expansions*	10
	Presentations: *POW 10: Growth of Rat Populations*	15
	POW 11: Twin Primes	10
	Vertex Form to Standard Form	50
	Homework: *How Much Can They Drink?*	5

Materials and Supplies

All IMP classrooms should have a set of standard supplies and equipment, and students are expected to have materials available for working at home on assignments and at school for classroom work. Lists of these standard supplies are included in the section "Materials and Supplies for the IMP Classroom" in *A Guide to IMP*. There is also a comprehensive list of materials for all units in Year 2.

Listed here are the supplies needed for this unit. General and activity-specific blackline masters are available for presentations on the overhead projector or for student worksheets. The masters are found in the *Fireworks* Unit Resources under Blackline Masters.

Fireworks

- No extra materials needed

More About Supplies

- Graph paper is a standard supply for IMP classrooms. Blackline masters of 1-Centimeter Graph Paper, ¼-Inch Graph Paper, and 1-Inch Graph Paper are provided so that you can make copies and transparencies for your classroom. (You'll find links to these masters in "Materials and Supplies for Year 2" in the Year 2 guide and in the Unit Resources for each unit.)

Assessing Progress

Fireworks concludes with two formal unit assessments. In addition, there are many opportunities for more informal, ongoing assessment throughout the unit. For more information about assessment and grading, including general information about the end-of-unit assessments and how to use them, consult the *Year 2: A Guide to IMP* resource.

End-of-Unit Assessments

This unit concludes with in-class and take-home assessments. The in-class assessment is intentionally short so that time pressures will not affect student performance. Students may use graphing calculators and their notes from previous work when they take the assessments.

Ongoing Assessment

Assessment is a component in providing the best possible ongoing instructional program for students. Ongoing assessment includes the daily work of determining how well students understand key ideas and what level of achievement they have attained in acquiring key skills.

Students' written and oral work provides many opportunities for teachers to gather this information. Here are some recommendations of written assignments and oral presentations to monitor especially carefully that will offer insight into student progress.

- *Using Vertex Form* will illustrate students' ability to pull together and use the various components of the vertex form of a quadratic.
- *Squares and Expansions* will demonstrate students' developing understanding of the technique of completing the square.
- *How Much Can They Drink?* will provide information on students' developing understanding of how to find the maximum value of a quadratic function to find the solution to a problem in context.
- *Another Rocket* will show how well students are prepared to address the unit problem.
- *A Fireworks Summary* is a reflective piece in which students summarize their work on the unit problem.
- *A Quadratic Summary* is a reflective piece in which students summarize their understanding of the big ideas of the unit.

Supplemental Activities

Fireworks contains a variety of activities at the end of the student pages that you can use to supplement the regular unit material. These activities fall roughly into two categories.

- **Reinforcements** increase students' understanding of and comfort with concepts, techniques, and methods that are discussed in class and are central to the unit.

- **Extensions** allow students to explore ideas beyond those presented in the unit, including generalizations and abstractions of ideas.

The supplemental activities are presented in the teacher's guide and the student book in the approximate sequence in which you might use them. Below are specific recommendations about how each activity might work within the unit. You may wish to use some of these activities, especially the later ones, after the unit is completed. In addition to these activities, you may want to use supplemental activities from Patterns that were not assigned or completed.

***What About One?* (reinforcement)** This problem-solving activity may be assigned anytime during in the unit. As in *POW 10: Growth of Rat Populations,* students have to be organized.

***Quadratic Symmetry* (extension)** Assign anytime after *Vertex Form For Parabolas*. Quadratic graphs are symmetrical about a vertical line through their vertex. In this activity, students use the general vertex form of a parabola to explore the relationship between the coordinates of corresponding points on either side of the line of symmetry.

***Subtracting Some Sums* (reinforcement)** You might use this activity with students who are having difficulty simplifying with negatives.

***Subtracting Some Differences* (reinforcement)** You might use this activity with students who are having difficulty simplifying with negatives.

***Choosing Your Intercepts* (reinforcement)** Assign anytime after *Crossing the Axis*. This activity gives students practice with finding an equation in vertex form for a parabola given the vertex and x-intercepts.

***A Lot of Symmetry* (reinforcement)** Assign anytime after *Distributing the Area II*. The special binomial product $(x + n)(x - n) = x^2 - n^2$ is introduced here in the context from *A Lot of Changing Sides*.

***Divisor Counting* (extension)** This problem-solving activity is best assigned after students have worked on *POW 11: Twin Primes*. The activity asks students to look for numbers that have a given number of divisors.

***The Locker Problem* (extension)** This classic problem makes a good follow-up to *Divisor Counting*.

***Equilateral Efficiency* (extension)** Assign this activity, which introduces Hero's formula for finding the area of a triangle in terms of its side lengths, after *Revisiting Leslie's Flowers*. With a little algebra, Hero's formula can be derived using the same approach used for finding the altitude of a triangle in *Revisiting Leslie's Flowers*.

***Check It Out!* (extension)** Use this activity with *Finding Vertices and Intercepts* or anytime during the unit. The activity introduces the notion that solving radical equations like $\sqrt{2x-3} = -5$ by squaring both sides may introduce extraneous roots.

***The Quadratic Formula* (extension)** Assign after *Coming Down*. The activity asks students to apply and then derive the quadratic formula. You may want to go through the derivation with them.

***Let's Factor Some More!* (reinforcement)** Assign after *Let's Factor!* This activity encourages students to use an area model to factor quadratics in which the coefficient of the x^2 term isn't 1.

***Vertex Form Challenge* (extension)** Assign after *Solve That Quadratic!* The activity gives students practice changing quadratic functions in standard form with leading coefficients other than 1 or –1 into vertex form. The problems involve fractions and decimals.

***A Big Enough Corral* (extension)** Assign after *Solve That Quadratic!* This activity explores quadratic inequalities. To do Question 2, students must be able to factor quadratics.

***Factors of Research* (extension)** Assign after *Solve That Quadratic!* The activity suggests further areas of exploration in the topic of factoring. Question 2 asks for a generalization of the difference of squares introduced in the supplemental activity *A Lot of Symmetry*.

***Make Your Own Intercepts* (extension)** Assign after *Quadratic Choices*. The activity builds on the idea that students can now easily find a quadratic that has two given x-intercepts. For example, for intercepts $x = 4$ and $x = 2$, the quadratic $y = (x - 4)(x - 2)$ will do. However, it isn't the *only* quadratic with those intercepts. All quadratics $y = a(x - 4)(x - 2)$ for any real number a will also work.

***Quadratic Challenges* (reinforcement)** Assign after *Quadratic Choices*. The activity offers three more quadratic equations for students to solve. (A graphing calculator would make finding the requested decimal answers too easy.)

***Standard Form, Factored Form, Vertex Form* (reinforcement)** Assign after *Quadratic Choices*. The activity pulls together the relationships among standard form, factored form, x-intercepts, vertex, and vertex form and makes a good group assignment.

© 2010 Interactive Mathematics Program

A Quadratic Rocket

Intent

In these activities, students are introduced to the unit problem and begin to explore the properties of the parabolic graphs of quadratic functions.

Mathematics

Victory Celebration introduces a classic problem in projectile motion in which a projectile travels along a path determined by the function $y = 160 + 92x - 16x^2$. This is a **quadratic function,** and its graph is a **parabola**. From an algebraic point of view, the key features of a parabola are

- its symmetry

- its turning point, called the *vertex*

- its intercepts

Other features, such as the focus and directrix, are important from a geometric point of view but are not studied in this unit.

In physics, quadratic functions are written with the constant first—in this case the height off the ground, 160 feet. In mathematics, quadratic functions are typically written with the x^2 term first. The physics format is used when referring to the unit problem, but the mathematics format is used elsewhere.

Students first explore the unit problem informally. Then they analyze the key features of parabolas and how they connect to the values of the parameters a, h, and k in the **vertex form** of the equation of a quadratic function, $y = a(x - h)^2 + k$.

Progression

The early activities introduce quadratic functions. The remaining activities ask students to make connections between the symbolic and graphical representations of these functions. In addition, students will begin work on the unit's first POW.

Victory Celebration

POW 10: Growth of Rat Populations

A Corral Variation

Parabolas and Equations I

Parabolas and Equations II

Rats in June

Parabolas and Equations III

Vertex Form for Parabolas

Using Vertex Form

Crossing the Axis

Is It a Homer?

Victory Celebration

Intent

This activity engages students in the unit problem. Students sketch the rocket situation, paraphrase the questions needed to solve the unit problem, and use a formula to find height values for specific time values.

Mathematics

The unit problem is a classic projectile-motion problem. By engaging in the unit problem in an informal way, students will explore some input and output numbers connected by the formula

$$h(t) = 160 + 92t - 16t^2$$

They will use the new vocabulary term **parabola** to describe the shape of the rocket's path and **quadratic function** to describe equations of the form $f(x) = ax^2 + bx + c$.

Progression

After reviewing the activity as a class, students work in groups on the questions. They then present their work, in the form of posters and oral reports, to the class.

Approximate Time

35 minutes

Classroom Organization

Whole class, then groups, followed by whole-class presentations

Doing the Activity

Have volunteers read the activity aloud. You might ask whether anyone has launched rockets in science classes or as a hobby.

Assign each group to prepare a poster displaying the information asked for in Questions 1 through 4.

Discussing and Debriefing the Activity

Have a few groups present the information on their posters to the class. As part of their work, they will probably have come up with steps like these:

- Find the value of t that makes $h(t)$ a maximum.

- Substitute this value for t in the function $h(t)$ to find the rocket's maximum height.

- To find out how long the rocket will be in the air, solve the equation $h(t) = 0$.

Have groups share their methods and results for Question 4. Students probably found (by graphing or guess-and-check) that the rocket will reach its maximum height in a little under 3 seconds and that the maximum height will be about 290 feet.

If students haven't yet graphed the height function, have them do so. Discuss how the graph confirms or corrects their ideas about the maximum height.

At which point on the graph is the function at its maximum? This means finding both coordinates of that point. At the maximum point in the rocket's path, the exact value of t is 2.875. (Students might not get exact answers.)

What does the landing time for the rocket mean in terms of the graph? Students likely found that the rocket would reach the ground after a little more than 7 seconds. This estimate involves interpreting the rocket's landing in terms of the function $h(t)$.

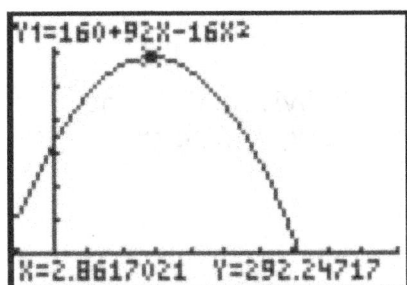

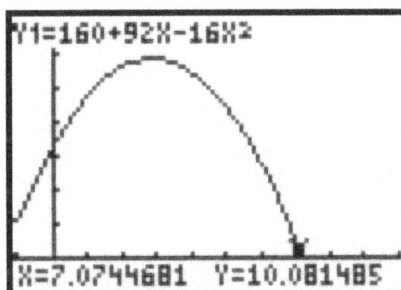

Students' explanations of these estimates will probably reflect that they tried to make $h(t)$ close to zero. Make this issue explicit by asking for an equation whose solution will give the time needed for the rocket to return to the ground. **What equation could you set up whose solution would tell you when the rocket hits the ground?** Students should see that trying to get $h(t)$ equal to zero is the same as solving the equation $160 + 92t - 16t^2 = 0$. (The exact value for the time for the rocket to land is irrational, so students cannot get the exact answer by graphing or guess-and-check.)

Key Questions

At which point on the graph is the function is at its maximum? This means finding both coordinates of that point.

What does the landing time for the rocket mean in terms of the graph?

What equation could you set up whose solution would tell you when the rocket hits the ground?

POW 10: Growth of Rat Populations

Intent

This POW provides an opportunity for students to organize data, devise solution methods, and describe and justify their work in writing.

Mathematics

In the problem, a female rat will produce a litter of six (three males and three females) every 40 days. Her female offspring will do the same, starting 120 days after their birth, and their female offspring will do the same. The time period in question, one year, allows for several generations of females to begin reproducing. The key mathematical skill students will encounter as they try to chart the growth of the rat population is the systematic organization and representation of data.

Progression

After a whole-class introduction, students work on the POW individually. In the upcoming activity *Rats in June,* they will partially solve this POW, finding the number of rats as of June 1.

Approximate Time

10 minutes for introduction

1–3 hours (at home)

20 minutes for presentations

Classroom Organization

Whole-class introduction with small-group discussion, then individuals, followed by whole-class presentations

Doing the Activity

You might have volunteers read the POW aloud and then give students about 10 minutes for discussion in groups. Schedule presentations for about seven days from today if you are on a 50-minute schedule.

Discussing and Debriefing the Activity

Though there is a unique, correct answer to this POW (1808), there are many ways to organize the data. After the presentations, give other students time to explain their organizational schemes. Focus the discussion on the process of working on the problem, without getting bogged down in the arithmetic.

Key Question

What were some schemes you used to keep the data organized?

Supplemental Activity

What About One? (reinforcement), a problem-solving activity that may be assigned anytime during the unit, requires students to be organized.

A Corral Variation

Intent

This activity presents another real-world situation involving a quadratic function, this one related to the corral problems in the unit *Do Bees Build It Best?*

Mathematics

In this classic area-maximization problem, students derive an expression for the area of the family of rectangles created by a fixed length of fence along three sides, with the fourth side provided by another long fence. Students will use this quadratic expression in one variable to calculate areas for several dimensions and then try to find the dimensions that enclose the maximum area.

Progression

After a brief introduction, students work on this activity individually and share results in a class discussion. The discussion should include how this problem relates to *Victory Celebration*.

Approximate Time

5 minutes for introduction

25 minutes for activity (in class or at home)

15 minutes for discussion

Classroom Organization

Individuals, followed by a whole-class discussion

Doing the Activity

Have students read through the activity. Answer any questions they have.

Discussing and Debriefing the Activity

For the discussion of Question 2, you might want to create a class table with headings "Length," "Width," and "Area" and collect answers.

For Question 3, there may be several versions of the area expression, including these:

$$x(500 - 2x) \qquad (500 - 2x)x \qquad 500x - 2x^2 \qquad 2[x(250 - x)]$$

For now, don't try to show these to be algebraically equivalent; the distributive property will be presented in *The Form of It All*. Instead, simply show that each gives the same output for any input.

Now explore with the class what the activities *Victory Celebration* and *A Corral Variation* have in common, in terms of both the graphs and the algebra. Here are some issues to address about the graphs of these problems.

- The graphs are not linear; instead, they go up and then down.

- The graphs each have a unique maximum. In both problems, a primary goal is to find this unique maximum, both its x- and y-coordinates. (In *Victory Celebration,* another major goal is to find the positive x-intercept.) Mention that some real-world problems might involve a minimum rather than a maximum, and draw a graph to illustrate a quadratic function with a minimum point. Identify this point on the graph, whether maximum or minimum, as the **vertex**. You might use the phrase "extreme value" to refer to this point generically, without specifying whether it is a maximum or minimum.

- The graphs are symmetrical. (For *Victory Celebration,* this may not be as clear, as students have probably looked only at the portion of the graph from $x = 0$ to the positive x-intercept.) You might ask students to illustrate how this symmetry works. They might point out that in each graph, there are generally two x-values for each y-value and that the two x-intercepts for the corral graph are such a "matched pair."

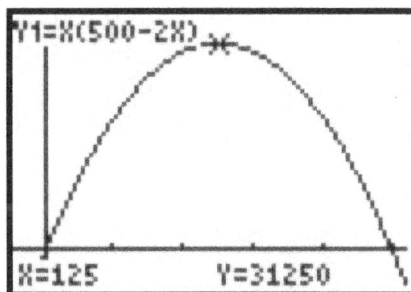

Here are some issues to address about the algebra of these problems.

- Both problems involve an expression with an x^2 term. As needed, multiply out the expressions students present to show that each has an x^2 term, reviewing that x^2 means simply $x \cdot x$ and that $x \cdot 2x$ is $2x^2$. (All that is needed is a general awareness that the product has an x^2 term. The topic of multiplying binomials will be treated more fully in upcoming activities.) Note that one way to get an x^2 term is by multiplying two x terms or, more generally, by multiplying two linear expressions.

- Identify an expression like these as a **quadratic expression** and the function it defines as a **quadratic function**. Without getting too formal, clarify that a quadratic has at most three terms: an x^2 term, perhaps an x term, and perhaps a constant term.

- Introduce the term **standard form** for a quadratic function written as $y = a x^2 + bx + c$.

Key Questions

What dimensions give the maximum area? What is the maximum area?

Parabolas and Equations I

Intent

This activity and those that follow introduce the term **parabola** and familiarize students with parabolic-shaped graphs and their properties.

Mathematics

In the sequence of activities *Parabolas and Equations I, II,* and *III,* students use calculators to explore what happens to the graph of $y = x^2$ when the parameters a, k, and h in $y = ax^2$, $y = x^2 + k$, and $y = (x - h)^2$ are changed. The graphing exploration culminates in *Vertex Form for Parabolas*, in which the ideas from all three activities are pulled together to present the general vertex form, $y = a(x - h)^2 + k$.

In this first activity, students explore the relationship between changes in the parameter a in $y = ax^2$ and corresponding changes in the graph.

Progression

Students work on the activity individually, with help from group mates as needed.

Approximate Time

20 minutes

Classroom Organization

Whole-class introduction, followed by individual work within small groups

Doing the Activity

You might begin by demonstrating the use of a graphing calculator to graph the rocket's height function, $h(x) = 160 + 92x - 16x^2$. Point out that this is one member of a family called **quadratic functions**. The general form of this family is $y = ax^2 + bx + c$, where a, b, and c represent numbers. Ask, What are the values of a, b, and c in the rocket function?

Have students reset their calculators to the standard viewing window, $-10 \leq x \leq 10$ and $-10 \leq y \leq 10$. Have them store in Y_1 the simplest member of this family, $Y_1 = x^2$. You can suggest they keep $Y_1 = x^2$ as they work and put other graphs in Y_2, Y_3, and so on, as done in the following illustrations. This makes it easier to compare each graph with the simplest parabola and address the question, With each change in a, what has changed in the graph?

```
Plot1 Plot2 Plot3
\Y1■X²
\Y2■2X²
\Y3■-.5X²
\Y4=
\Y5=
\Y6=
\Y7=
```

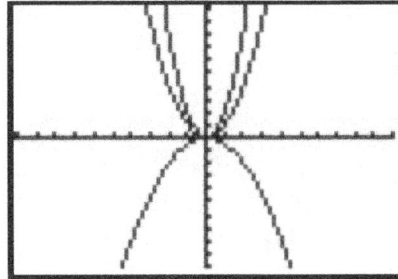

Circulate as students work, answering questions about calculator operation and reminding students to record what they find, including making sketches of the graphs. Make sure students are working through the material on their own, as using a graphing calculator is a basic mathematics skill that can be mastered only by practice.

Discussing and Debriefing the Activity

During the discussion, introduce the idea of *concavity*. Help students relate the terms *concave up* and *concave down* to the graphs and the equations.

If time allows, you might ask students to create original designs of their own using no more than five equations, including both linear and quadratic functions.

Key Questions

What are the values of a, b, and c in the rocket function?

With each change in a, what has changed in the graph?

Parabolas and Equations II

Intent

This activity continues the exploration begun in *Parabolas and Equations I*, in which students familiarized themselves with parabolic graphs.

Mathematics

In this second activity of the sequence *Parabolas and Equations I, II,* and *III,* students explore the relationship between changes in the parameter k in $y = x^2 + k$ and corresponding changes in the graph.

Progression

After successfully completing *Parabolas and Equations I,* students should begin work on this activity individually, with help from group mates as needed.

Approximate Time

15 minutes

Classroom Organization

Individuals within small groups

Doing the Activity

Remind students to reset their calculators to the standard viewing window, $-10 \le x \le 10$ and $-10 \le y \le 10$.

Discussing and Debriefing the Activity

After most students have completed this activity, call the class together and have volunteers share their answers to Question 2. There is more than one way to make each design.

Then say, **Now we know how *a* and *k* change the graphs of $y = ax^2$ and $y = x^2 + k$. Put these two equations together, and we have $y = ax^2 + k$.**

Ask students to predict, without actually graphing, how the graph of each of the following equations would compare with the simplest graph, $y = x^2$.

$$y = 2x^2 + 3$$
$$y = -2x^2 - 3$$
$$y = 0.5x^2 - 4$$
$$y = -0.01x^2 + 5$$

Have students check their predictions by graphing the equations.

Rats in June

Intent

This activity will clarify some of the issues in *POW 10: Growth of Rat Populations,* as well as assure that students get started on the problem.

Mathematics

The key mathematical skill students will encounter as they try to chart the growth of the rat population is the systematic organization and representation of data.

Progression

In this activity, students will partially solve the POW, finding the number of rats as of June 1.

Approximate Time

40 minutes for activity (at home or in class)

10 minutes for discussion

Classroom Organization

Individuals, then groups, followed by whole-class discussion

Doing the Activity

Tell students that this activity will get them started on the POW and that as they work they should jot down any questions they have about the problem.

Discussing and Debriefing the Activity

Give students a few minutes to share results in their groups. Then ask, How many rats, male and female, are there as of June 1?

The goal of this discussion is to clear up any misinterpretations of the problem. In particular, students should see that the original female will have four litters by June 1 and that the females in the first litter will themselves each have a litter by that date. (As of June 1, there are 22 female and 22 male rats.)

Parabolas and Equations III

Intent

This activity continues the work from *Parabolas and Equations I* and *II*, in which students familiarized themselves with parabolic graphs.

Mathematics

In this third activity in the sequence *Parabolas and Equations I, II,* and *III*, students explore the relationship between changes in the parameter h in $y = (x - h)^2$ and corresponding changes in the graph.

Progression

After successfully completing *Parabolas and Equations I* and *II,* students should begin work on this activity individually, with help from group mates as needed. In the next activity, *Vertex Form for Parabolas*, the ideas from these three activities are pulled together when students work with vertex form, $y = a(x - h)^2 + k$.

Approximate Time

15 minutes

Classroom Organization

Individuals within small groups

Doing the Activity

Once again, remind students to reset their viewing windows to standard.

Discussing and Debriefing the Activity

After most students have completed this final activity in the sequence, call the class together and have volunteers share their answers to Question 2.

Then say, **Now we know how a, k, and h change the graphs of $y = ax^2$, $y = x^2 + k$, and $y = (x - h)^2$. Put these three equations together, and we have $y = a(x - h)^2 + k$.**

Ask students to predict, without actually graphing, how the graph of each of the equations here would compare with the simplest graph, $y = x^2$.

$$y = 2(x - 4)^2 + 3$$
$$y = -2(x - 4)^2 - 3$$
$$y = 0.5(x - 2)^2 - 4$$
$$y = -0.01(x + 2)^2 + 5$$

Have students check their predictions by graphing the equations.

If time allows, you might ask students to create another original design using no more than five equations, including both linear and quadratic functions.

Vertex Form for Parabolas

Intent

In this activity, the ideas from *Parabolas and Equations I, II,* and *III* are pulled together as students work with the general vertex form for quadratics, $y = a(x - h)^2 + k$.

Mathematics

The **vertex form** for quadratics, $y = a(x - h)^2 + k$, allows one to treat a parabola as a transformation of the "parent" graph $y = x^2$. The vertex form gives the parabola's vertex at the point (h, k). If $a > 0$, the parabola will be concave up; if $a < 0$, the parabola will be concave down. The absolute value of a will determine how vertically stretched or compressed the graph looks in the viewing window.

The activities in *A Quadratic Rocket* focus only on the relationship between the parameters in vertex form and the corresponding graphs. Later in the unit, students will learn how to convert quadratics in standard form, $y = ax^2 + bx + c$, into vertex form by completing the square.

Progression

After an opening class discussion, students work individually on this activity to re-create graphical designs.

Approximate Time

25 minutes

Classroom Organization

Whole class, then individuals within small groups

Doing the Activity

To introduce the activity, you might lead the class in a brief discussion.

We know how a, k, and h change the graphs of $y = ax^2$, $y = x^2 + k$, and $y = (x - h)^2$. Put these three equations together, and we have $y = a(x - h)^2 + k$.

Discussing and Debriefing the Activity

Have students share their equations in their groups. Then have each group choose one set of equations for each design and display them for presenting to the class. As solutions are presented, ask the class to comment on how closely the graphs of each set of equations match the given designs.

Although many answers are possible for each design, in Questions 1 and 2 the values of the parameters should reflect the symmetries in the graphs. For example, in Question 1, the vertex values for the parabolas on the upper left and upper right should have the same y-coordinate and the x-coordinates should be negatives of each other. The equations $y_1 = (x + 3)^2 + 2$, $y_2 = x^2$, and $y_3 = (x - 3)^2 + 2$ will work, with vertices at $(-3, 2)$, $(0, 0)$, and $(3, 2)$.

Supplemental Activity

Quadratic Symmetry (extension) further explores quadratic graphs, which are symmetrical about a vertical line through their vertex. Students use the general vertex form of a parabola to explore the relationship between the coordinates of corresponding points on either side of the line of symmetry.

Using Vertex Form

Intent

Students apply the ideas developed in the last few activities about the graphical effects of the parameters in the vertex form of a quadratic equation.

Mathematics

The vertex form of the equation of a quadratic function, $y = a(x - h)^2 + k$, allows one to determine the vertex of the graph, the direction of the graph's concavity, and the relative amount of vertical stretch or compression of the parabola. In this activity, students derive equations to produce a particular graphical design and read the vertex coordinates from several equations.

Progression

Students work individually on the activity and discuss their findings in class.

Approximate Time

25 minutes for activity (at home or in class)

15 minutes for discussion

Classroom Organization

Individuals, followed by whole-class discussion

Doing the Activity

This activity requires little or no introduction.

Discussing and Debriefing the Activity

Have several students or groups present their solutions to Question 3. The problem was to be done by estimating the coordinates of the vertex from the graph. The Zoom feature can be helpful in accurately estimating these values.

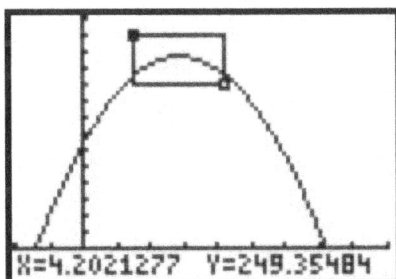

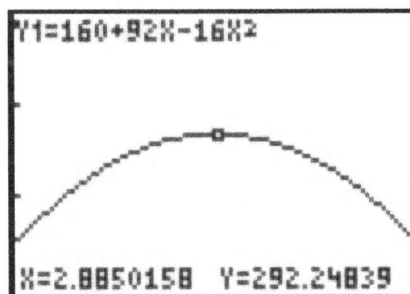

How might you write the rocket equation in vertex form? Based on the estimate for the vertex shown in the previous graph, students might propose an equation like

$$y = -(x - 2.88)^2 + 292$$

This will produce a parabola with the correct vertex and pointing in the right direction, but that still needs to be stretched vertically (by increasing $|a|$) to match the graph of $y = 160 + 92x - 16x^2$.

Crossing the Axis

Intent

In this activity, students begin to develop an understanding of the processes through which they might find the x-intercepts of the graph of a quadratic function.

Mathematics

In this activity, students will use knowledge about the location of the vertex and whether the value of a in $y = a(x - h)^2 + k$ is positive or negative (indicating the graph's concavity) to determine the number of x-intercepts a parabola has. In general, there are three cases: one, two, or no x-intercepts.

Students will also find the equation for a parabola given the vertex and one x-intercept. The values of a parabola's vertex determine the parameters h and k in the equation $y = a(x - h)^2 + k$. If, in addition, an x-intercept (or "zero") of the parabola is given, the value of the parameter a can be found by substituting 0 for y and the x-intercept value for x and solving.

Progression

Students work on the activity in groups and then share their work with the whole class.

Approximate Time

25 minutes

Classroom Organization

Groups, followed by whole-class discussion

Doing the Activity

Students may need help with the logic for Questions 1 to 4. You may also want to work through the example preceding Question 5 with the whole class.

Discussing and Debriefing the Activity

Elicit answers for Questions 1 to 4. The goal is for students to recognize that the location of the vertex and the concavity of the graph determine the number of x-intercepts the graph of a quadratic function has.

Have volunteers share their work for Questions 5 and 6. Students might use the approach modeled in the student book to find the value of the parameter a.

Supplemental Activities

Subtracting Some Sums (reinforcement) gives students practice simplifying with negatives.

Subtracting Some Differences (reinforcement) gives students practice simplifying with negatives.

Choosing Your Intercepts (reinforcement) offers students practice with finding an equation in vertex form for a parabola given the vertex and *x*-intercepts.

Is It a Homer?

Intent

Students apply ideas from *Using Vertex Form* and *Crossing the Axis* in a real-world context.

Mathematics

In this activity, students treat the path of a baseball as a parabola, which allows them to assume that the flight of the ball is symmetrical: its highest point occurs halfway along its path from the ground up to that maximum height and back to the ground again. Given the location of this maximum point and one x-intercept, and treating the ground as the x-axis, students will use this information to find an equation for the parabolic path. Then they will use the equation to find a particular output (the height of the ball) for a given input (the location of the center field fence). Finally, they will determine whether the ball will clear the 15-foot fence.

Progression

Students work individually on this activity and discuss their findings in class. The discussion will also refer to their work on the previous exploratory activities.

Approximate Time

10 minutes for introduction

20 minutes for activity (at home or in class)

15 minutes for discussion

Classroom Organization

Individuals, followed by whole-class discussion

Doing the Activity

You might suggest that students read the activity and sketch the parabolic path of the ball from home plate.

Consider home plate to be (0, 0) with the x-axis running along the ground directly beneath the path of the ball and the y-axis indicating the height of the ball above the ground. Some students may want to place (0, 0) about two feet above home plate, where the bat meets the ball. Either positioning is fine.

Discussing and Debriefing the Activity

Have individuals or groups present their solutions. As needed, ask questions such as these listed here.

What are the coordinates of the vertex of the ball's parabolic path?

How did you find an equation for the ball's path?

Once you have an equation, how can you decide whether the ball will clear the fence for a home run?

The function is approximately $y = -0.002(x - 200)^2 + 80$. When $x = 380$, the value of y is 15.2. So, will the ball clear a 15-foot fence? Yes. In fact, because the batter will actually hit the ball from above home plate, the ball would clear the fence by an even greater margin. Note that these calculations do not take into consideration the effect of any wind.

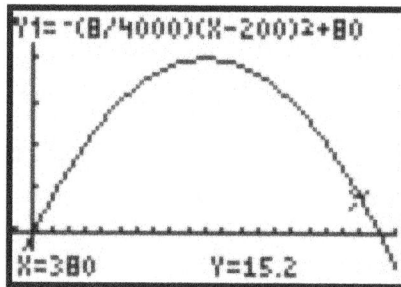

You might complete this discussion by asking students to make up a similar problem for another context, such as a soccer kick at goal or a football kick for a field goal.

Key Questions

What are the coordinates of the vertex of the ball's parabolic path?

How did you find an equation for the ball's path?

Once you have an equation, how can you decide whether the ball will clear the fence for a home run?

The Form of It All

Intent

These activities focus on the symbolic representation of quadratic functions as well as some of the algebraic skills needed to multiply and factor quadratic expressions.

Mathematics

The distributive property is at the heart of the "doing and undoing" skills developed in *The Form of It All*. Students will use an area model to connect multiplication of polynomials to multiplication of numbers and to develop procedures for multiplying expressions of the forms $(x + a)(x + b)$ and $(x + c)^2$. For example, to multiply $(2x + 5)(x + 3)$,

$$
\begin{array}{r}
2x + 5 \\
\times \quad x + 3 \\
\hline
15 \\
6x \\
5x \\
+ \quad 2x^2 \\
\hline
2x^2 + 11x + 15
\end{array}
$$

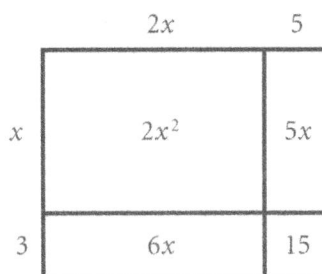

	$2x$	5
x	$2x^2$	$5x$
3	$6x$	15

Students will use these techniques to convert quadratic functions from vertex form to standard form.

Progression

The activities begin with a review of the distributive property in a numeric context and then extend it to work with polynomials. The area model is introduced as a tool for multiplying polynomials and completing the square. Students move back and forth between graphical and symbolic representations of quadratic functions. In addition, they present their solutions to the unit's first POW and begin work on the second.

A Lot of Changing Sides

Distributing the Area I

Views of the Distributive Property

Distributing the Area II

Square It!

Squares and Expansions

POW 11: Twin Primes

Vertex Form to Standard Form

How Much Can They Drink?

A Lot of Changing Sides

Intent

This activity introduces the area model for multiplying binomials.

Mathematics

The Form of It All emphasizes building meaning into the symbolic manipulations associated with multiplying and factoring polynomials. The key idea here is the distributive property of multiplication over addition: $a(b + c) = ab + ac$. Given the natural connection between multiplication and the area of a rectangle, this property is easily represented in this manner:

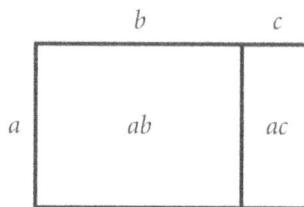

Total area: $a(b + c) = ab + ac$

This activity provides students with this geometric model for the process of multiplying binomials (and other polynomials). In this activity, students will use symbols to express the areas of rectangles created by altering squares with sides of length X. As the unit develops, they will also use this area model for completing the square and for factoring.

Progression

Students work as a whole class on Question 1 and then tackle the remaining five questions in their groups.

Approximate Time

30 minutes

Classroom Organization

Whole-class introduction, followed by small groups

Doing the Activity

If students mention that this activity seems to be a digression from work with vertex form, you can suggest that knowing how to multiply two binomials is necessary to transform quadratics written in vertex form, like $y = (x - 4)^2 - 7$, into an equivalent equation in standard form, like $y = x^2 - 8x + 9$.

Have students read the introduction and Question 1. Then ask them to explain what, exactly, they are being asked to do. Be sure they identify the two steps of drawing a sketch and finding an expression for the new area.

To get them started, ask them to sketch and label the original lot. They should get something simple like this.

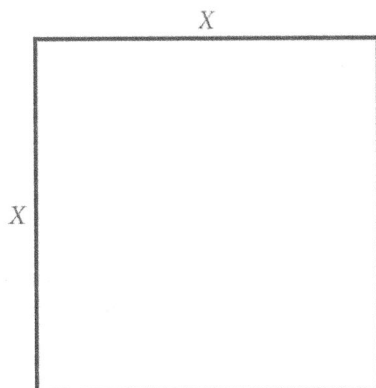

Then have them discuss in their groups how to sketch and label the lot described in Question 1. They should come up with something like the next diagram, in which the shaded area represents the "new" portion of the lot.

Ask groups to come up with two expressions for the area of the new lot, as described in the activity. Students should see that the area is the product $(X + 3)(X + 4)$. For the second expression, you may have to suggest that they divide the diagram into rectangles. As the next diagram shows, the area is also the sum $X^2 + 4X + 3X + 12$. Students do not need to combine terms, but it's okay if they do.

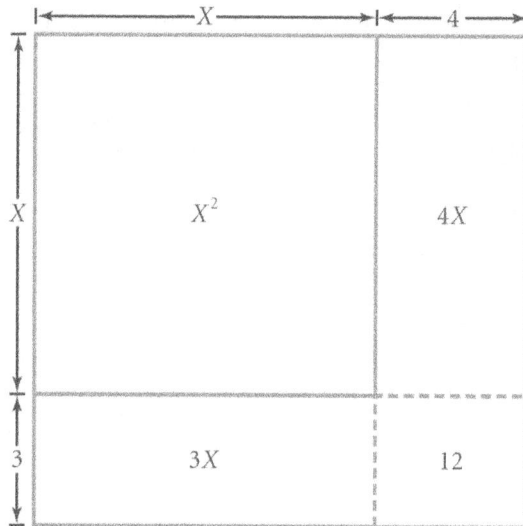

It is important that the expression without parentheses comes from the diagram as a sum of areas, even if some students know how to multiply out the product $(X + 3)(X + 4)$.

Have students work in their groups on the rest of the activity.

Discussing and Debriefing the Activity

The key goal is for students to see the connection between the area expressed as a sum of separate rectangles and as the product of the new length and width. For example, the area in Question 2 might be expressed as the sum $X^2 + 5X$ or as the product of the length and width, $X(X + 5)$. You might ask students, What is the term for two algebraic expressions that represent the same thing? If necessary, remind students that these are called *equivalent expressions*.

You can have students check the equivalence of their expressions by substituting numeric values. They can also check by graphing the functions defined by the equations, such as $Y = X^2 + 4X + 3X + 12$ and $Y = (X + 3)(X + 4)$ for Question 1. They should see that the graphs are identical. Be sure they understand that neither of these "checks" proves the equivalence.

The area model falls apart somewhat when it comes to dealing with negative numbers, as in Questions 5 and 6, but students may have some creative ways of dealing with "negative area." Most students are comfortable with a diagram like the next one for Question 5, even though it shows both negative lengths and negative areas.

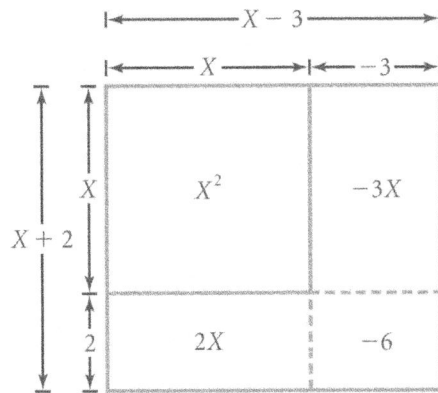

Another way students might draw the situation in Question 5 is shown here.

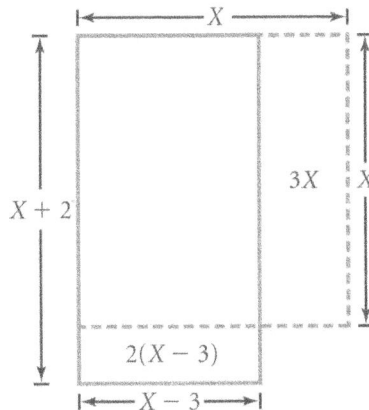

In this case, the two ways to describe the area symbolically are

$$(X + 2)(X - 3) \quad \text{and} \quad X^2 - 3X + 2(X - 3)$$

In the second expression, students begin with the original square's area of X^2, subtract the rectangle with area $3X$, and add the rectangle with area $2(X - 3)$. Question 6 can be modeled geometrically using these same ideas, though it is a bit trickier.

If time permits, you may want to do some more examples like those in *A Lot of Changing Sides,* focusing on the issue of signs when one or both of the constant terms is negative.

Distributing the Area I

Intent

Students apply the area model to the multiplication of numeric and algebraic expressions.

Mathematics

This activity lays the foundation for understanding the area model for multiplying binomials. Students will multiply the binomials $(a + b)$ and $(c + d)$, using specific values for the variables, to find areas. In some cases, they will be given values for area and must work backward to find the factors whose product gives that area. Working backward, or "undoing," prepares students to understand factoring with polynomials.

Progression

Students work individually or in groups on the activity and share results in a brief class discussion.

Approximate Time

5 minutes for introduction

15 minutes for activity (at home or in class)

15 minutes for discussion

Classroom Organization

Individuals, followed by whole-class discussion

Doing the Activity

Review the use of the area model, presented in the introduction to this activity in the student book, for multiplying algebraic expressions.

Discussing and Debriefing the Activity

Have students or groups present their solutions.

Question 1 is a straightforward task of labeling an area model and computing areas.

Starting with Question 2, students need to reason to find missing lengths and areas. In Question 2, if area IV is 6 square units and length d is 2, then length b must be 3. The question being addressed is, If I know the area of a rectangle, what must the dimensions be? This is analogous to the question, If I know the product of two numbers, what must the numbers be? Students are "undoing" multiplication, or factoring.

Question 6 is challenging and involves some guess-and-check. If the total area is 864 square units, the overall dimensions of the rectangle must be factors of 864. Students know that $a = 30$ and $c = 20$, so they need to find lengths b and d such that the three remaining areas add to 264 square units. There are only two whole-number factor pairs of 864 with one number greater than 30 and the other greater than 20: 32 and 27, and 36 and 24. So, there are two possible answers: $b = 2$ and $d = 7$, or $b = 6$ and $d = 4$.

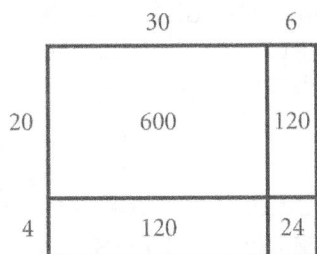

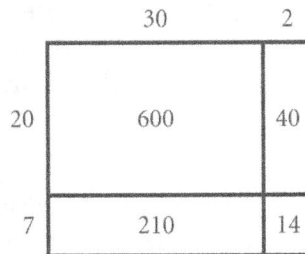

	30	6
20	600	120
4	120	24

	30	2
20	600	40
7	210	14

Views of the Distributive Property

Intent

This activity gives students some ways to think about the distributive property in arithmetic, both in terms of familiar algorithms and using an area model. Students then apply these ideas to the multiplication of algebraic expressions.

Mathematics

The distributive property for multiplication over addition is written symbolically as $a(x + y) = ax + ay$. This property is more generally applied to the multiplication of polynomials as follows:

$$(a + b)(c + d) = (a + b)(c) + (a + b)(d) = ac + bc + ad + bd$$

In this activity, students are asked to consider that they have been using this property since they learned how to multiply two multidigit numbers. Once the familiar algorithm is expanded to show (using the area model) the partial products and how they are combined to form the product, students are asked to generalize this approach to algebraic expressions. For example,

To multiply 32 • 94,	To multiply (2x + 5)(x + 3),
30 + 2	2x + 5
× 90 + 4	× x + 3
8	15
120	6x
180	5x
+ 2700	+ 2x²
3008	2x² + 11x + 15

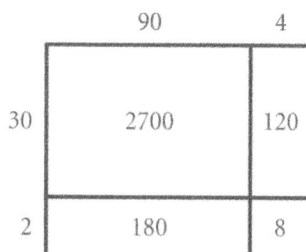

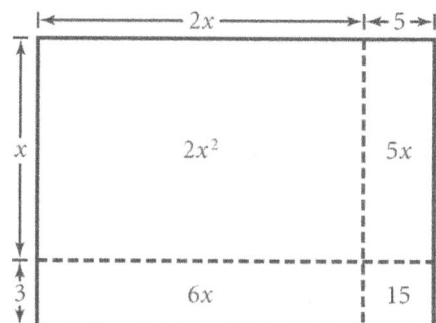

Progression

Students read the activity and then work on the tasks in groups, with a follow-up discussion focused on arithmetic–algebra connections and combining terms.

Approximate Time

35 minutes

Classroom Organization

Groups, followed by whole-class discussion

Doing the Activity

Have students read the expository material in the student book and then work on the questions in their groups.

Discussing and Debriefing the Activity

Although Questions 1 and 2 may seem elementary, they serve as a valuable foundation for upcoming work with variables. In discussing the long form of multiplication, you may need to review some basics of place value so students see where the zeros are coming from. You might also bring out that the standard multiplication algorithm (the "short form") involves combining some terms mentally. This is indicated in the student book, but may be worth repeating.

For Question 3, focus on the facts that the multiplication involves six partial products and that the corresponding area model has six separate rectangles.

Students are likely to have different methods for Question 4a (preferably including at least one long-form and one short-form method) and various ideas about how (or whether) to line up the terms. For instance, they might produce either of these forms.

$$
\begin{array}{r}
2x + 5 \\
\times\ \ x + 3 \\
\hline
15 \\
6x \\
5x \\
+\ \ \ \ 2x^2 \\
\hline
\end{array}
\qquad \text{or} \qquad
\begin{array}{r}
2x + 5 \\
\times\ \ \ \ \ x + 3 \\
\hline
15 \\
6x \\
5x \\
+\ \ \ 2x^2 \\
\hline
\end{array}
$$

Others may follow a pattern that looks more like the standard multiplication algorithm.

$$
\begin{array}{r}
2x + 5 \\
\times\ \ \ \ x + 3 \\
\hline
6x + 15 \\
+\ 2x^2 + 5x \\
\hline
\end{array}
\qquad \text{or} \qquad
\begin{array}{r}
2x + 5 \\
\times\ \ \ \ \ \ \ \ x + 3 \\
\hline
6x + 15 \\
+\ 2x^2 + 5x \\
\hline
\end{array}
$$

Emphasize that because the purpose of a written format is to make the work as clear and as easy to do as possible, people may prefer different methods.

The diagram for Question 4b should look something like this.

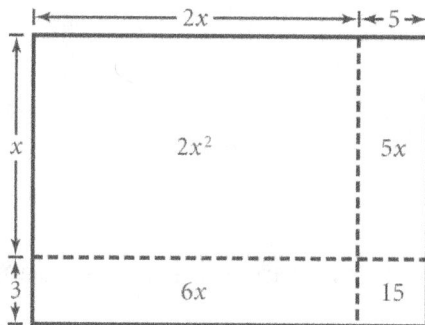

If needed, point out the analogy between this diagram and the diagram in Question 2, which involves numeric multiplication, including the connection between the rectangles in the diagram and the partial products in the long form of the multiplication. Also reinforce the idea that a product of two binomials will have four terms, although it's likely that two of them can be combined.

Combining Terms

Whatever method students use for getting partial products or partial sums, they need to know how to combine the terms into a single final expression. Review the idea of combining like terms, perhaps using the analogy of place value to bring out that multiples of x are like 10s, multiples of x^2 are like 100s, and so on.

Ask, **In general, how do you multiply two sums? What are the partial products? Where do the partial products come from?** Students should be able to articulate that the partial products are all the products obtained by multiplying a term of one sum by a term of the other sum.

Have students explain the process using an area model. Emphasize the value of being able to go back and forth between the symbol manipulation and the area model for multiplication. The goal is for students to understand this multiplication process, rather than simply memorizing a rule such as FOIL (*first, outside, inside, last*).

You may want to review multiplication examples in which one or both factors involve subtraction, such as $(x + 3)(x - 2)$ or $(x - 4)(x - 6)$. You may also want to have students work out an example in which each factor has more than two terms, such as $(x + y + 3)(2x + 3y + 1)$.

Distributing the Area II

Intent

This activity continues the themes of using an area model to represent multiplication and factoring of numbers, and connecting arithmetic multiplication to polynomial multiplication.

Mathematics

Students return to the area model introduced in *Distributing the Area I* and use it to represent multiplication of two binomials and to find the binomials that produce a quadratic product. They also continue to use the vertical form introduced in *Views of the Distributive Property* to multiply two polynomials.

Progression

Students work on the activity individually. The subsequent class discussion will clear up any lingering questions and emphasize the arithmetic-algebra connection.

Approximate Time

20 minutes for activity (at home or in class)

15 minutes for discussion

Classroom Organization

Individuals, then groups, followed by whole-class discussion

Doing the Activity

No introduction is necessary for this activity.

Discussing and Debriefing the Activity

In Question 4, students are asked for the first time to "undo" a quadratic expression that is the answer to a multiplication problem. Rather than showing them a procedure for factoring, this is a chance for students to use the model they have been developing to reason their way through the process.

If the total area is $x^2 + 18x + 80$ and the area has four parts, students might begin by labeling a rectangle this way.

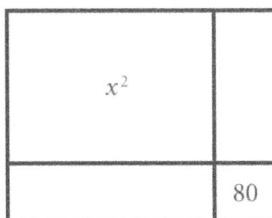

From this diagram, they can reason that the dimensions of the upper-left square are x and x and the dimensions of the lower-right square are two numbers whose

product is 80. In addition, the two factors of 80 must be numbers that will create remaining areas that add to $18x$. The area model can then be completed as follows.

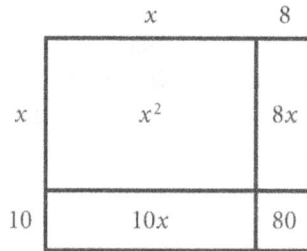

	x	8
x	x^2	$8x$
10	$10x$	80

When multiplying polynomials, students should simplify their final answers by combining like terms. Traditionally, terms are written in order of descending powers of the variable.

Supplemental Activity

A Lot of Symmetry (reinforcement) introduces the special binomial product $(x + n)(x - n) = x^2 - n^2$ in the context from *A Lot of Changing Sides*.

Square It!

Intent

Students apply the multiplication skills they have been developing to the task of rewriting quadratic functions from vertex form into standard form.

Mathematics

Students focus on rewriting quadratic functions in vertex form, $y = a(x - h)^2 + k$, in standard form, $y = ax^2 + bx + c$. First they focus on using an area model to square binomials. After rewriting functions in standard form, they combine this work with previous content to use the vertex and intercepts of the graph of a parabola to write its vertex-form equation and then rewrite the equation in standard form.

Progression

Students work on the activity in groups and discuss their findings as a whole class.

Approximate Time

25 minutes

Classroom Organization

Groups, followed by whole-class discussion

Doing the Activity

Tell students that they will now use what they have learned about multiplying binomials to change quadratics written in vertex form into standard form.

Discussing and Debriefing the Activity

As you discuss Question 1, you might have students draw area models to illustrate what's going on and use the diagrams to explain any apparent patterns. For instance, they might note that they always get two same-area rectangles. In the diagram here for Question 1a, these are the lower-left and upper-right rectangles. This leads to the observation that the coefficient of x in the square of a binomial is twice the constant term of the expression being squared with the same sign as that constant term. Similarly, the area of the lower-right corner is the square of the constant term of the expression being squared, and this value is the constant term in the result.

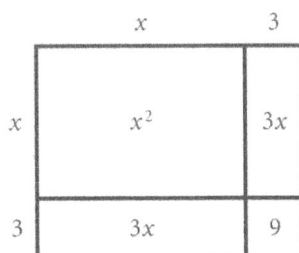

	x	3
x	x^2	$3x$
3	$3x$	9

You might want to reassure students that they can use the area model even if the constant term is negative. For instance, the next diagram shows that $(x - 5)^2$ is equal to $x^2 - 10x + 25$.

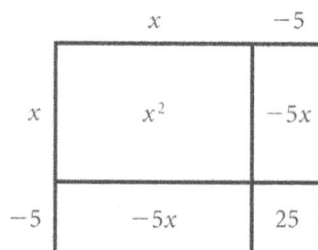

You might have students compare this area approach with the process of expanding the expressions in Question 1 using the vertical-multiplication form from *Distributing the Area II* and bring out again that these products are illustrative of the distributive property.

Question 2a is straightforward, but students may have questions about 2b and 2c. These entail three steps, designed to follow the order of operations.

- Square the binomial.

- Apply the distributive property, multiplying each term of the resulting trinomial by the given coefficient (in Question 2b by 3 and in Question 2c by 0.5).

- Add like terms and write the complete answer.

In Question 3, to find an equation from a graph, students need to estimate the coordinates of the vertex and at least one x-intercept and then use the techniques from *Crossing the Axis* and *Is it a Homer?* to find the equation in vertex form. They should check their answers by graphing their functions and comparing them with the pictured graphs.

Squares and Expansions

Intent

Students use the area model for multiplication to begin to develop, in a meaningful way, a familiar symbolic procedure.

Mathematics

In this activity, students try the first step of the technique traditionally called **completing the square,** which later will be applied to changing standard form into vertex form and to deriving the quadratic formula. Using an area model, students recognize that they are literally "completing a square" when they create a perfect square. For example, given the expression $x^2 + 10x$, adding 25 completes the square, resulting in the expression $x^2 + 10x + 25 = (x + 5)^2$.

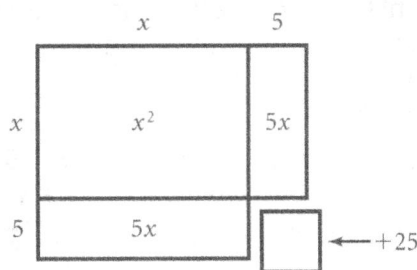

Progression

Students will work on the three parts of this activity—completing the square, expanding expressions, and sketching a graph—individually and share their results in a class discussion.

Approximate Time

10 minutes for introduction

25 minutes for activity (at home or in class)

15 minutes for discussion

Classroom Organization

Individuals, followed by whole-class discussion

Doing the Activity

Read the activity and work through Questions 1a and 1b as a class.

Discussing and Debriefing the Activity

Focus the discussion of Question 1 on the use of area models to explain the process. As each diagram is presented, be sure students write the corresponding expression as the square of a linear expression. For instance, they might get a preliminary diagram like this one for Question 1d and thus need to add 4 to get a perfect square. They can then write the expression as $(x - 2)^2$.

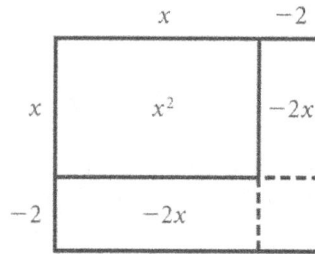

You may want to help students articulate that in each part of Question 1, the constant term of the binomial being squared is half the coefficient of x in the original expression. For instance, -2, from $x - 2$, is half the coefficient -4 in $x^2 - 4x$.

As students present their work for Question 2, you might focus their attention on being careful about signs.

Question 3 will remind students of the usefulness of vertex form in sketching graphs. They should immediately realize from the equation that the vertex is at (5, 10) and that this point is a maximum, so the graph opens downward. To sketch the graph, the only further information they need is a rough idea of where the x-intercepts are. They should reason that they want $(x - 5)^2 = 10$, so $x - 5 = \pm\sqrt{10}$. That is, $x - 5$ is about 3 or -3, so x is approximately 8 or 2. This is a good opportunity to get students to articulate the symmetry principle for parabolas once again.

POW 11: Twin Primes

Intent

In exploring and then justifying an interesting property of the number system, students will also be applying some of the skills they are developing in this unit, including the manipulation of algebraic expressions.

Mathematics

Twin primes are prime numbers that are 2 units apart, such as 5 and 7, 41 and 43, 71 and 73, 137 and 139, and 281 and 283. This POW centers on an interesting property of twin primes. If you add 1 to their product, the result is a perfect square and a multiple of 36. In this activity, students look for twin primes, explore this property, and then try to prove that it is always true.

Progression

After a brief introduction, students will work on this problem individually outside of class and then submit detailed write-ups of their solutions. Presentations will follow.

Approximate Time

10 minutes for introduction

1–3 hours (at home)

20 minutes for presentations

Classroom Organization

Whole class, then individuals, followed by whole-class presentations

Doing the Activity

To introduce this POW, review the definition of *twin prime*. To be sure students understand the properties being presented, ask them to find all the twin primes between 50 and 100.

If students have done any calculator programming, you may want to suggest they write a program that will give all twin primes between two input numbers.

Discussing and Debriefing the Activity

The POW asks students to use a variable and prove that multiplying twin primes and adding 1 results in a perfect square. If students use x and $x + 2$ to represent a pair of twin primes, the task is then to show that $x(x + 2) + 1$ is a perfect square. As students may have discovered, this expression is a perfect square for any integer x, because it is equal to $(x + 1)^2$.

If some students got this algebraic representation but did not succeed in showing that the expression is also a multiple of 36, you might ask, **If a perfect square is a multiple of 36, what is true about the square root of that number?** This might lead them to recognize that they need to show that $x + 1$ (which is the number between the two twin primes) is a multiple of 6.

Showing that the result is a multiple of 36 builds on the fact that if a prime greater than 3 is divided by 6, the remainder must be either 1 or 5. In symbolic terms, this means that a prime is of the form $6k + 1$ or $6k + 5$.

Supplemental Activities

Divisor Counting (extension) asks students to look for numbers that have a given number of divisors. This problem-solving activity is best assigned after students have worked on *POW 11: Twin Primes*.

The Locker Problem (extension) is a classic problem that makes a good follow-up to *Divisor Counting*.

Vertex Form to Standard Form

Intent

This activity gives students more practice with algebraic manipulation and parabolas in another real-world context.

Mathematics

One definition of a **parabola** is the set of points that are equidistant from a fixed line and a fixed point not on that line. The fixed line is called the *directrix,* and the fixed point is called the *focus* of the parabola. One important property of parabolas is that they concentrate incoming energy—light or sound, for example—at the focus. This reflective property is the real-world context for Question 9, in which students are to find, graph, and then use the equation of a parabolic mirror with its vertex at the origin and passing through the point (40, 10).

Progression

Students will work on this activity individually and share ideas in a class discussion.

Approximate Time

5 minutes for introduction

25 minutes for activity (at home or in class)

40 minutes for discussion

Classroom Organization

Individuals, followed by whole-class discussion

Doing the Activity

To introduce this activity, you might mention some of the types of objects that have parabolic cross sections, such as automobile headlights, TV dishes, makeup mirrors, and sound amplifiers. You might also suggest that students may want attempt to build a solar cooker out of cardboard parabolic slices and aluminum foil.

Discussing and Debriefing the Activity

You might have groups prepare presentations on each of Questions 1 to 8.

Then discuss how students approached Question 9. The only way to get the length for the cardboard support is to find the equation for a parabola with vertex at (0, 0) and using the point (40, 10) to find the value of *a*.

How Much Can They Drink?

Intent

This final activity of *The Form of It All* revisits a context first encountered in the unit *Do Bees Build It Best?* in which students look for the vertex of a quadratic function to find the dimensions of a prism with the maximum possible volume.

Mathematics

In *A Corral Variation*, students learned that the area of a rectangle formed by a fixed length L of fence along three sides is given by $A = x(L - 2x)$. This area function's graph is a parabola, and its vertex can be used to find the maximum possible area.

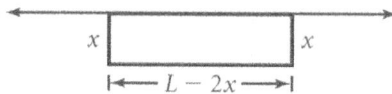

In *Do Bees Build It Best?* students learned that the volume of a prism is the product of the area of the prism's base and the prism's height. They now encounter a watering trough with a rectangular cross section formed just as the rectangular fence above and are asked to find the dimensions of the trough that will give the maximum volume.

Progression

Students work on this activity individually and discuss their results in class.

Approximate Time

10 minutes for introduction

20 minutes for activity (at home or in class)

15 minutes for discussion

Classroom Organization

Whole class, then individuals, followed by whole-class discussion

Doing the Activity

You might want to review the idea that the volume of a rectangular solid is the product of the solid's length, width, and height by posing some volume problems.

How many cubes 1 inch on an edge will a box measuring 5 inches by 7 inches by 10 inches hold?

A wave table in the physics lab is an open-top box with a rectangular base. It holds 9996 cubic centimeters and is 42 centimeters long and 28 centimeters wide. What is its height?

A farmer has 36 feet of fencing and wants to enclose the maximum rectangular area for his llamas. Find the dimensions of three possible areas he could enclose. What do you think the maximum area is? Why?

Discussing and Debriefing the Activity

You might begin the discussion with Question 3 and return to Questions 1 and 2 if needed to illustrate the process of finding the volume.

Some students may need help developing and making sense of the expression $40 - 2x$ for the width. You might have them make a table of specific examples, including the one in the activity with $x = 5$. Once they have this expression, it should be straightforward to get a formula for the volume equivalent to $V = 80(40 - 2x)x$.

Because the volume expression as given here involves three factors, including the constant 80, students might not recognize that they are working with a quadratic function. Have them multiply out the expression and rearrange the terms (getting $-160x^2 + 3200x$) to emphasize this. Seeing the expression as a quadratic will allow students to connect "finding the maximum" with the idea of finding the vertex of the corresponding graph.

Students might find the vertex (and thus find the maximum volume) by a guess-and-check approach or by tracing on their calculator graphs. They might also use the symmetry of the parabola and conclude that the x-coordinate of the vertex is halfway between x-coordinates of the x-intercepts. Once they have the x-coordinate of the vertex, they can get the volume by substitution.

Bring out that students can find the x-intercepts easily by tracing. But they can also find them symbolically, because the function is expressed as a product of linear terms. The product is zero whenever one of the factors is zero, namely, when $x = 0$ or $x = 20$. If needed, write the explicit step $40 - 2x = 0$ and go from that to $x = 20$.

To foreshadow later work, point out that the factored form of the volume equation makes finding the x-intercepts easy and also allows one to find the vertex using the symmetry.

Putting Quadratics to Use

Intent

These activities focus on problems that can be solved using a quadratic function in vertex form and the algebra skills needed to transform quadratic functions into vertex form.

Mathematics

In *The Form of It All,* students modeled the multiplication of polynomials using the area of a rectangle. The activities in *Putting Quadratics to Use* build on that work and on the general emphasis on the "doing and undoing" aspect of algebraic thinking present throughout the curriculum. Students learn how to undo the process of transforming quadratic functions in vertex form into standard form, using the method typically known as "completing the square."

Students also continue to derive quadratic functions for a variety of real-world contexts, some of which they have encountered in earlier units. With their developing skill of transforming functions into vertex form, students find and interpret key points on the related graphs, including vertices and intercepts.

Progression

In *Putting Quadratics to Use,* students work with six contexts that lead to quadratic functions. By the end of these activities, they will have had many opportunities to transform quadratic functions back and forth between standard form and vertex form, and to use vertex form to maximize or minimize functions and solve quadratic equations.

Revisiting Leslie's Flowers

Emergency at Sea

Here Comes Vertex Form

Finding Vertices and Intercepts

Another Rocket

Profiting from Widgets

Pens and Corrals in Vertex Form

Vertex Form Continued

Revisiting Leslie's Flowers

Intent

This activity uses a context first encountered in the unit *Do Bees Build It Best?* as an opportunity for students to apply their new skills for working with quadratic equations.

Mathematics

In *Leslie's Fertile Flowers*, students applied the Pythagorean theorem for right triangles twice to find the dimensions and area of a triangular flower bed. In this activity, students will write and solve the quadratic equation they derive symbolically by solving each of the equations below for h^2 and setting the two expressions equal.

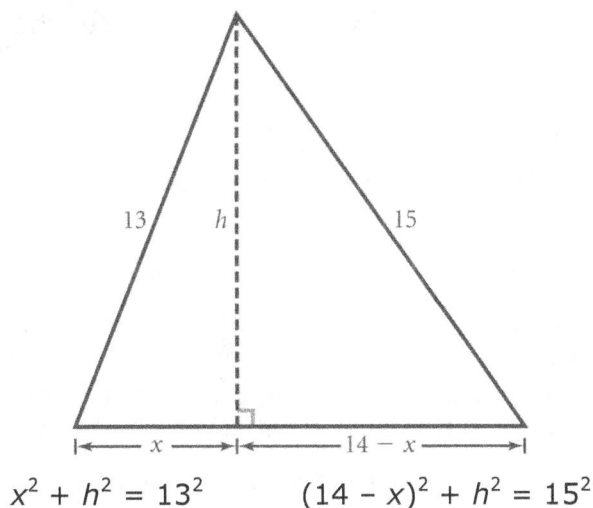

$$x^2 + h^2 = 13^2 \qquad (14 - x)^2 + h^2 = 15^2$$

Progression

Students work on this activity in groups.

Approximate Time

30 minutes

Classroom Organization

Groups, followed by whole-class discussion

Materials

Students' notes from the unit *Do Bees Build it Best?*

Doing the Activity

This activity uses the situation presented in *Leslie's Fertile Flowers,* from *Do Bees Build It Best?* In that activity, students found the value of *x* by a guess-and-check approach. They did this by applying the Pythagorean theorem to the two right triangles created by the altitude of the triangle. The goal now is for students to work through the algebra needed to solve the equation directly. You might choose to work through Question 1 as a class, both to review the situation and to help students set up the algebraic representation.

One key step is using $14 - x$ to represent the second portion of the 14-foot side. Letting h represent the altitude, the left-hand triangle then leads to the equation $13^2 - x^2 = h^2$ while the right-hand triangle leads to the equation $15^2 - (14 - x)^2 = h^2$. Combining the two gives the one-variable equation

$$13^2 - x^2 = 15^2 - (14 - x)^2$$

Before setting students off on the challenge of solving this equation, you might point out that expanding the expression on the right side is precisely the skill they have been developing.

You might also note that the two expressions each represent the vertex form for a parabola, so solving this equation is like finding the place where two parabolas intersect.

Discussing and Debriefing the Activity

The first stage of the solution will likely be to remove the parentheses, probably getting

$$169 - x^2 = 225 - 196 + 28x - x^2$$

One issue that may arise is how to deal with the x^2 terms. Although the mechanics are the same as with linear terms, students may be less comfortable with a quadratic expression. It may be worth taking a moment to discuss why "adding x^2 to both sides" is a legitimate action that yields an equivalent equation.

In one sequence of steps or another (and you may want to get more than one sequence), students should be able to simplify the equation to something like $140 = 28x$ and see that $x = 5$.

Have students confirm this answer in at least a couple of ways. One confirmation is to check that this value satisfies the original version of the equation, $13^2 - x^2 = 15^2 - (14 - x)^2$. Another way is to verify that splitting the side of length 14 into parts of length 5 and 9 gives the same result for the length of h. As $13^2 - 5^2$ and $15^2 - 9^2$ are both equal to 144, both give $h = 12$.

Supplemental Activity

Equilateral Efficiency (extension) introduces Hero's formula for finding the area of a triangle given the lengths of its sides. With a little algebra, Hero's formula can be derived with the same approach used for finding the altitude of the triangle in *Revisiting Leslie's Flowers*.

Emergency at Sea

Intent

This activity gives students another opportunity to set up and solve a quadratic equation for a real-world context.

Mathematics

As in *Revisiting Leslie's Flowers*, students will use the Pythagorean theorem on the two right triangles created by dividing a scalene triangle along an internal altitude. The solution to the resulting quadratic equation will give the length of the third side of the triangle, and substitution will give the triangle's height.

Progression

Students work on this activity individually and share their results with the class.

Approximate Time

5 minutes for introduction

20 minutes for activity (at home or in class)

15 minutes for discussion

Classroom Organization

Individuals, followed by whole-class discussion

Doing the Activity

Read the activity as a class, making sure students understand the diagram.

Discussing and Debriefing the Activity

If students go through a process similar to that used for *Revisiting Leslie's Flowers*, they might create a diagram like the one here, in which $y = 1000 - x$. This will likely lead to the equation $550^2 - x^2 = 800^2 - (1000 - x)^2$.

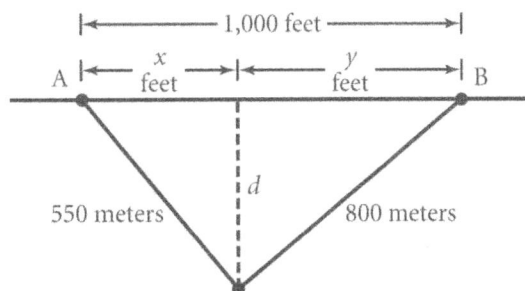

Students should see that their recent work expanding expressions with parentheses is directly applicable here. Have them apply what they have learned to get a parentheses-free expression equivalent to $(1000 - x)^2$. Once they have done so, the equation will probably look like this.

$$550^2 - x^2 = 800^2 - (1{,}000{,}000 - 2000x + x^2)$$

You may need to point out that the entire expression for $(1000 - x)^2$, namely, $1{,}000{,}000 - 2000x + x^2$, is being subtracted, which is indicated by enclosing the expression in parentheses.

When 550^2 and 800^2 are replaced by their numeric values, the equation becomes

$$302{,}500 - x^2 = 640{,}000 - (1{,}000{,}000 - 2000x + x^2)$$

The equation simplifies as follows:

$$302{,}500 - x^2 = 640{,}000 - 1{,}000{,}000 + 2000x - x^2$$
$$302{,}500 = 640{,}000 - 1{,}000{,}000 + 2000x$$
$$302{,}500 = -360{,}000 + 2000x$$
$$302{,}500 + 360{,}000 = 2000x$$
$$662{,}500 = 2000x$$
$$x = 331.25 \text{ feet (distance to tower A)}$$

Thus the distance to tower B is 668.75 feet. For Question 4, students can find the distance d from the boat to the shore from the equation $d^2 = 550^2 - 331.25^2$, which gives $d \approx 439.06$, and then verify that this value also satisfies the equation $d^2 + 668.75^2 = 800^2$.

Here Comes Vertex Form

Intent

This activity emphasizes reversing, or undoing, the process of changing a quadratic function from vertex form to standard form.

Mathematics

Students use the process of completing the square to transform quadratic expressions in standard form, $ax^2 + bx + c$, into vertex form. To simplify things, the expressions all have a leading coefficient of 1 or –1. The groundwork for this process was laid in *Squares and Expansions*, where students learned that expressions of the form $x^2 + kx$ can be made into a "square," geometrically and symbolically, by adding the constant term $\left(\dfrac{k}{2}\right)^2$.

Progression

Students work on the activity individually or in groups. In Questions 1 to 4, they change quadratic functions from standard form to vertex form and interpret the results. In Questions 5 and 6, they are given functions in vertex form and are asked to find their x-intercepts.

Approximate Time

30 minutes

Classroom Organization

Groups or individuals, followed by whole-class discussion

Doing the Activity

You may need to demonstrate the technique of completing the square. The idea underlying the process seems obvious: adding 0 to any quantity leaves the quantity's value unchanged. But 0 can take many forms, such as 5 – 5 and –3.65 + 3.65. There are many approaches to the mechanics of this process, and it may be helpful for students to see a variety of methods.

Discussing and Debriefing the Activity

Have volunteers share their methods for transforming each equation. Here are two ways students might approach the expression $x^2 + 4x + 1$ from Question 1, for example.

- Begin by putting the constant term in front to get $1 + (x^2 + 4x)$. Then look at $x^2 + 4x$. Completing the square for this expression requires adding 4, but if we add 4, we must also subtract 4 (for a net change of 0), so we write $x^2 + 4x$ as $x^2 + 4x + 4 - 4$. The original expression thus becomes $1 + (x^2 + 4x + 4) - 4$, which can then be written in vertex form as $(x + 2)^2 - 3$.

- Look at the initial two terms of the expression, $x^2 + 4x$, and think about how to complete the square. The expression needs a constant term of 4 to complete the square and obtain $(x + 2)^2$. The original expression, $x^2 + 4x + 1$, has a constant term of 1 and is thus "3 short," so we rewrite it as $(x + 2)^2 - 3$. Similarly, in Question 2, the constant term of 15 is "6 too much," and the vertex form is $(x - 3)^2 + 6$.

Whatever method students use, you might have them substitute one or two values for x to verify that the vertex form they get is equivalent to the original expression. This is also another opportunity to look at how to get the vertex from the vertex form.

In Questions 3 and 4, students need to exercise care to get the signs correct. For Question 3, they might begin by writing the equation as $y = 3 - (x^2 - 8x)$, as in the first method described earlier, or as $-(x^2 - 8x - 3)$, based on the second method, and proceed with the expression in parentheses as follows:

- The expression in parentheses, $x^2 - 8x$, needs 16 to be a perfect square, so we can write it as $x^2 - 8x + 16 - 16$ or, equivalently, $(x - 4)^2 - 16$. So the overall expression becomes $3 - [(x - 4)^2 - 16]$, which is equivalent to $3 - (x - 4)^2 + 16$. This simplifies, in vertex form, to $-(x - 4)^2 + 19$.

- Looking at the expression in parentheses, $x^2 - 8x - 3$, we see that the initial portion, $x^2 - 8x$, needs 16 to be a perfect square, $(x - 4)^2$. The expression has a constant term of -3, which is "19 short" of 16, so the expression in parentheses can be thought of as $(x - 4)^2 - 19$. The original expression is therefore $-[(x - 4)^2 - 19]$, which is equivalent to $-(x - 4)^2 + 19$ in vertex form.

The next activity, *Finding Vertices and Intercepts*, gives further practice with completing the square. Students will have more opportunities to work with the process later, so it is not necessary for them to master the skill of turning quadratics into vertex form at this time.

Finding Vertices and Intercepts

Intent

Students transform functions into vertex form and use the result to find vertices and x-intercepts.

Mathematics

One advantage of having a quadratic function in vertex form is being able to find graphical features like the coordinates of the vertex and the direction of concavity by inspection. Another advantage, emphasized here, is being able to use this form to solve a quadratic equation symbolically. For example, if the vertex form of a quadratic function is $y = (x + 2)^2 - 4$, it is relatively easy to find the x-value(s) when $y = 0$, if such solutions exist: $(x + 2)^2 - 4 = 0$ is equivalent to $(x + 2)^2 = 4$, so $x + 2 = \pm 2$, and $x = -4$ or 0.

Progression

Students work individually on the activity and share their findings in a class discussion.

Approximate Time

5 minutes for introduction

20 minutes for activity (at home or in class)

15 minutes for discussion

Classroom Organization

Individuals, followed by whole-class discussion

Doing the Activity

This is a good time to review two key ideas.

- "Finding an x-intercept" means making y equal to zero.
- The two square roots of a given number are found by setting up and solving two separate equations.

Discussing and Debriefing the Activity

Questions 1 and 3 both have whole-number x-intercepts.

In Question 1, the vertex form is $y = (x + 2)^2 - 1$, so students want $(x + 2)^2 - 1 = 0$ or, equivalently, $(x + 2)^2 = 1$. Splitting this into two cases, $x + 2 = 1$ and $x + 2 = -1$, gives $x = -1$ and $x = -3$.

You might take this opportunity to bring out that the vertex has an x-coordinate of -2, which is the midpoint between the intercepts at $x = -1$ and $x = -3$, thus reviewing the idea that parabolas are symmetric about a vertical line drawn through the vertex. The two x-intercepts are 1 unit on either side from -2.

For Question 2, the vertex form is $y = (x - 3)^2 + 3$, and students should see that there are no intercepts. You might ask them to explain what's going on with the graph. For instance, they might say the graph "starts" at $y = 3$ and goes up from there because the expression $(x - 3)^2$ cannot be negative.

Question 4 presents the first case in which students are solving to get irrational intercepts. The vertex form is $y = -(x + 1)^2 + 5$, so students want $(x + 1)^2 = 5$.

Students might get an intuitive sense of how this works if they begin with a decimal approximation for $\sqrt{5}$. They can conclude from $(x + 1)^2 = 5$ that $x + 1$ is approximately either 2.24 or –2.24, giving the approximations $x + 1 \approx 2.24$ and $x + 1 \approx -2.24$, so $x \approx 1.24$ or –3.24.

Then have them look at how they can do something similar using the radical $\sqrt{5}$ to see that $x = \sqrt{5} - 1$ or $-\sqrt{5} - 1$. They can verify that these expressions are consistent with the earlier approximations. You might bring out that even in this case, the x-coordinate of the vertex is halfway between the two x-intercepts.

Supplemental Activity

Check It Out! (extension) introduces the notion that solving radical equations such as $\sqrt{2x - 3} = -5$ by squaring both sides may introduce extraneous roots. You can use this activity with *Finding Vertices and Intercepts* or anytime during the unit.

Another Rocket

Intent

Students use their developing methods for solving quadratic equations in the context of the unit problem, with a somewhat simpler function.

Mathematics

Students use the technique of completing the square once again to find the vertex and x-intercepts of a quadratic function. However, this time, the coefficient of the x^2 term is not 1.

Progression

Students work in groups on the activity. Initial questions about how to proceed with functions in which the leading coefficient is not 1 might lead to a class discussion.

Approximate Time

30 minutes

Classroom Organization

Groups

Doing the Activity

Allow students to begin without any preliminary discussion. Some will quickly notice that the coefficient of x^2 is not 1, so this problem is different from those they have been doing.

To answer their questions, you may want to do an example with the class. For instance, consider the task of putting $y = 2x^2 + 12x + 6$ into vertex form. There are various approaches for taking into account the coefficient 2 (from $2x^2$), each with advantages. Here are two such options.

- Factor out 2, getting $2(x^2 + 6x + 3)$, and proceed with the expression in parentheses as in one of the earlier methods to get $(x + 3)^2 - 6$. As a final step, simplify $2[(x + 3)^2 - 6]$ into vertex form.

- Isolate the constant term, writing the expression as $6 + 2(x^2 + 6x)$. Then write $x^2 + 6x$ as $(x + 3)^2 - 9$, and proceed to simplify $6 + 2[(x + 3)^2 - 9]$ into vertex form.

The first approach may be more effective for demonstrating that any quadratic expression can be put into vertex form: factor out the coefficient, put the rest into vertex form, and then undo the original factoring. The second approach sets up the expression initially to look more like vertex form and then only requires completing the square and dealing with the additional constant term.

Discussing and Debriefing the Activity

Once groups get a solution, they may ask you whether it is correct. Rather than answering directly, you might instead ask them to graph the original function on their calculators and then use the Zoom feature to estimate the x-intercepts. If their answers are correct, they should agree with the calculator values.

Later in the unit, in *Fireworks Height Revisited,* students will put the equation $h(t) = 160 + 92t - 16t^2$ from the unit problem in *Victory Celebration* into vertex form.

Profiting from Widgets

Intent

Students are introduced to another application of quadratic equations and are asked to find the maximum value of a quadratic function.

Mathematics

This activity presents an economics context for the mathematical work of the unit. Given an expression for the number of "widgets" sold as a linear function of price, students are to derive a revenue function, in which revenue is found by multiplying the number of widgets sold by the price per widget. This revenue function is quadratic, and students can use the skills they have been developing to find its vertex.

Progression

Students work on the activity individually and compare methods and results in a class discussion.

Approximate Time

5 minutes for introduction

20 minutes for activity (at home or in class)

15 minutes for discussion

Classroom Organization

Individuals or groups, followed by whole-class discussion

Doing the Activity

Have students read the activity in their groups and discuss what kind of equation they might use to represent the sales revenue.

Discussing and Debriefing the Activity

Begin by having students explain the equation for sales revenue, which should look something like

$$R = d(1000 - 5d)$$

You might give the class additional examples similar to Question 1 to analyze whether students are clear about the meaning and basis of this formula.

Students might have found the maximum revenue by various means, including guess-and-check and estimating from a graph. Another approach is to find the two x-intercepts ($d = 0$ and $d = 200$) and use the fact that the d-coordinate of the vertex is the midpoint of the d-intercepts. Because the expression is in factored form, students may find it easy to get the intercepts.

If students didn't do so on their own, ask them explicitly to find the maximum by putting the revenue function in vertex form. Here is a possible sequence of steps.

$$R = d(1000 - 5d)$$
$$= 1000d - 5d^2$$
$$= -5(d^2 - 200d)$$
$$= -5(d^2 - 200d + 10{,}000 - 10{,}000)$$
$$= -5[(d - 100)^2 - 10{,}000]$$
$$= -5(d - 100)^2 + 50{,}000$$

This final form gives the vertex as (100, 50,000).

Ask students to interpret this information in terms of the context. Acme will make the maximum possible profit by charging $100 per widget. At that price, they will sell 500 widgets and make $50,000 in revenue.

For Question 4, students should recognize that the marketing director's formula has flaws, such as that if the price is set above $200 per widget, the formula says the company will sell a negative number of widgets.

Pens and Corrals in Vertex Form

Intent

Students revisit two contexts from earlier in the curriculum, in which solutions were found informally, and solve them again using the more formal methods of this unit.

Mathematics

The two problem contexts in this activity come from *A Corral Variation* and from *Don't Fence Me In* in the unit *Do Bees Build It Best?*, in which students found that the rectangle with the greatest area for a given perimeter is a square. Students are asked to prove the correctness of their previous solutions using algebra. In each case, they will write a quadratic function, transform it into vertex form, and read the vertex coordinates to find the maximum value of the function.

Progression

Students work on the activity in groups and share their results in a class discussion.

Approximate Time

35 minutes

Classroom Organization

Groups, followed by whole-class discussion

Doing the Activity

This activity requires little or no introduction.

Discussing and Debriefing the Activity

For Question 1, ask for volunteers to present the steps of putting the expression into vertex form and how to use it to find the maximum area. In the steps shown here, the term added to create a perfect square is left as 125^2 until the last step. This technique might help students focus on the big picture rather than on the arithmetic.

$$
\begin{aligned}
y &= x(500 - 2x) \\
&= -2x^2 + 500x \\
&= -2(x^2 - 250x) \\
&= -2(x^2 - 250x + 125^2 - 125^2) \\
&= -2[(x - 125)^2 - 125^2] \\
&= -2(x - 125)2 + 2 \cdot 1252 \\
&= -2(x - 125)^2 + 31{,}250
\end{aligned}
$$

Connect this algebra with the real-world context by asking about the units for x and y (feet and square feet).

Students might prefer to justify the answer here based on the symmetry of the graph, which tells them that the vertex is halfway between the intercepts at $x = 0$ and $x = 250$. If so, use this as the stimulus for a discussion about the advantages and disadvantages of different methods. You might point out that the symmetry approach works well here because the function is in factored form, but that this is not the case for the rocket situation.

The algebra for Question 2 is even simpler, because the coefficient of x^2 is -1 instead of -2. The work might look like this.

$$y = x(100 - x)$$
$$= -x^2 + 100x$$
$$= -(x^2 - 100x)$$
$$= -(x^2 - 100x + 50^2 - 50^2)$$
$$= -[(x - 50)^2 - 50^2]$$
$$= -(x - 50)^2 + 50^2$$
$$= -(x - 50)^2 + 2500$$

If students find the "second side" of the rectangle as $\frac{200 - 2}{2}$ (subtracting the two sides of length x from the total perimeter of 200 and then dividing by 2 to get the other sides), point out that this expression is equivalent to $100 - x$.

By now students should be comfortable explaining, based on this final expression, that the maximum area occurs when x is 50 meters, so the rectangle is a square with an area of 2500 square meters.

Vertex Form Continued

Intent

This activity offers students additional experience with the types of problems they have been solving throughout *Putting Quadratics to Use*.

Mathematics

In this activity, students will transform functions in standard form into vertex form and vice versa. They will also find the symbolic representation of a quadratic function from its vertex and an additional point. Students should be growing more fluent with manipulating quadratic expressions. At this point, they are not expected to have mastered the process of transforming quadratics from standard form into vertex form when the coefficient of x^2 is other than 1 or −1. They should be able to work through such transformations as a group, though, as they did in activities like *Another Rocket*.

Progression

Students work on this activity individually and share results in a class discussion.

Approximate Time

20 minutes for activity (at home or in class)

15 minutes for discussion

Classroom Organization

Individuals, followed by whole-class discussion

Doing the Activity

This activity requires little or no introduction.

Discussing and Debriefing the Activity

You might follow up the work on Question 1 by asking about the *x*-intercepts of these quadratic functions. Ask whether students see a way to get the intercepts for Question 1b without putting the functions into vertex form. They should see that the function in Question 1c does not have any *x*-intercepts.

Question 2 provides practice with squaring binomials and manipulating expressions involving parentheses. You may want to encourage students to use intermediate steps, such first writing the expression in Question 2a as

$$-2(x^2 - 4x + 4) + 4$$

and then, after removing the parentheses, as

$$-2x^2 + 8x - 8 + 4$$

You may want to point out that this is an application of the distributive property. Also be on the alert for sign errors.

Questions 3 and 4 give more practice with finding an equation given the vertex and a point. You might ask students to rewrite each function in standard form.

For Question 4, the arc shape of the cross section of the cable forming the main span of the Golden Gate Bridge is actually a catenary curve, but a parabola shape is very close. The equation for the arc, using (0, 0) as the vertex and passing through the point (2100, 500), is $y = (500/2100)^2 \cdot x^2$. The height at the halfway point, where $x = 1050$, is $y = 125$.

Back to Bayside High

Intent

In *Back to Bayside High,* students return to and solve the unit problem.

Mathematics

These activities allow students to pull together all the ideas that have been at play throughout the unit:

- graphical and symbolic representations of quadratic functions, including connections between the vertex form of a function and key features of the graph, such as its vertex and direction of concavity

- properties of parabolic graphs, including symmetry and the vertex

- symbolic manipulations, many based on applications of the distributive property

Specifically, students will be transforming the function in the unit problem into vertex form and then using this form to find and interpret the graph's vertex and *x*-intercepts.

Progression

Students will solve the unit problem in the first two activities. In the final activity, they will summarize their solutions and methods in writing.

Fireworks in the Sky

Coming Down

A Fireworks Summary

Fireworks in the Sky

Intent

Students are now ready to return to the unit problem, presented in *Victory Celebration.* As they will see, this activity considers only part of the problem—the questions relating to the rocket's highest point. The question of when the rocket hits the ground is examined in the next activity, *Coming Down.*

Mathematics

Armed with the skills needed to transform a function in standard form into vertex form, students are now ready to tackle the unit problem. They will find the exact maximum height of the soccer team's rocket, and the exact time at which this height is reached, by reading the vertex coordinates from the vertex form of the function. The algebra of putting the expression $160 + 92t - 16t^2$ into vertex form is similar to students' work in *Another Rocket,* but the numbers make it somewhat more difficult.

Progression

Students work on the activity in groups. The class discussion that follows will focus on the procedures they used, as well as why they used them and why they work.

Approximate Time

30 minutes

Classroom Organization

Groups, followed by whole-class discussion

Doing the Activity

Before groups embark on this activity, you might remind them that they estimated the maximum height of the rocket at the beginning of the unit and emphasize that now they are to find it exactly, using vertex form.

Discussing and Debriefing the Activity

If some groups worked through the algebra, let them give presentations, but move slowly to be sure they don't lose the rest of the class. If students were not able to get all the details on their own, you may want to lead them through it so they can see the process in full. Ask volunteers to provide the ideas for individual steps as the whole class contributes by checking the mechanics.

Here are two possible sequences of steps for the algebraic transformation of the expression into vertex form, based on the second approach in the discussion of *Another Rocket.* Other sequences of steps are also possible. The steps in the two methods are the same except that the first uses fractions, while the second uses decimals. The steps are numbered for easy reference in the subsequent discussion.

If students use a decimal approach, emphasize that they are not to round off the decimals. (The fact that the denominator of the original fraction is a power of 2 implies it has a finite decimal expansion, so it's feasible to keep the values precise.)

Description	Fraction representation
Step 1: Begin with the original expression.	$160 + 92t - 16t^2$
Step 2: Isolate the constant term.	$160 - (16t^2 - 92t)$
Step 3: Factor out the coefficient of t^2 and simplify the fraction $\dfrac{92}{16}$.	$160 - 16\left(t^2 - \dfrac{23}{4}t\right)$
Step 4: Add and subtract to complete the square.	$160 - 16\left[t^2 - \dfrac{23}{4}t + \left(\dfrac{23}{8}\right)^2 - \left(\dfrac{23}{8}\right)^2\right]$
Step 5: Write the perfect square as the square of a binomial.	$160 - 16\left[\left(t - \dfrac{23}{8}\right)^2 - \left(\dfrac{23}{8}\right)^2\right]$
Step 6: Multiply through by 16.	$160 - 16\left(t - \dfrac{23}{8}\right)^2 + 16 \bullet \left(\dfrac{23}{8}\right)^2$
Step 7: Combine the constant terms.	$-16\left(t - \dfrac{23}{8}\right)^2 + \dfrac{1169}{4}$

The same process, using decimals, looks like this.

Description	Decimal Representation
Step 1: Begin with the original expression.	$160 + 92t - 16t^2$
Step 2: Isolate the constant term.	$160 - (16t^2 - 92t)$
Step 3: Factor out the coefficient of t^2 and write $92 \div 16$ as 5.75.	$160 - 16(t^2 - 5.75t)$
Step 4: Add and subtract to complete the square.	$160 - 16(t^2 - 5.75t + 2.875^2 - 2.875^2)$
Step 5: Write the perfect square as the square of a binomial.	$160 - 16[(t - 2.875)^2 - 2.875^2]$
Step 6: Multiply through by 16.	$160 - 16(t - 2.875)^2 + 16 \bullet 2.875^2$
Step 7: Combine the constant terms.	$-16(t - 2.875)^2 + 292.25$

If you need to lead the class through much of this algebra, here are some questions you might ask.

- Steps 2 and 3: **What initial steps did you use in other problems, especially when the coefficient of the squared term wasn't 1?**

- Step 4: **What needs to be added to complete the square?** Remind students, if needed, to subtract in order to compensate for adding. Whether students work with fractions or decimals, they might benefit from leaving the expression as the square of some number, either $\left(\dfrac{23}{8}\right)^2$ or 2.875^2, rather than finding the numeric value of the expression.

Even if they have difficulty working out all the steps, students should be able to follow the general process, to describe what's going on, and to use the final result.

Once students have the vertex form, you might want to give groups time to think about how to use it to answer the questions about the rocket. You can then have students compare the exact answers they get with the approximate solutions found at the beginning of the unit.

Here are the answers to this part of the unit problem.

- The rocket reaches its maximum height exactly $\dfrac{23}{8}$ seconds, or $2\dfrac{7}{8}$ seconds, or 2.875 seconds after launch.

- At its highest point, the rocket is exactly $\dfrac{1169}{4}$ feet, or $292\dfrac{1}{4}$ feet, or 292.25 feet above the ground.

Coming Down

Intent

This activity completes work on the unit problem begun in *Fireworks in the Sky,* as students find the time it takes for the rocket to return to the ground.

Mathematics

Having transformed the quadratic function in the unit problem from standard form to vertex form, students are now ready to find the exact amount of time it will take the rocket to return to the ground. To do this, they will set the vertex form of the function equal to zero and solve for *t.*

Progression

Students work on the activity individually and then participate in a whole-class discussion of the process of solving the equation.

Approximate Time

5 minutes for introduction

20 minutes for activity (at home or in class)

25 minutes for discussion

Classroom Organization

Individuals, followed by whole-class discussion

Doing the Activity

Read through the activity as a class. Make sure students have a general idea of the process for solving the problem.

Discussing and Debriefing the Activity

After students give an articulate statement of the connection between the algebra and the context, you may want to focus the discussion on the broad outline of the steps needed to get from vertex form to the *x*-intercept. These are the key steps for both the decimal and fractional forms of the function.

Description	Decimal representation	Fraction representation
Set up the equation.	$-16(t - 2.875)^2 + 292.25 = 0$	$-16\left(t - \dfrac{23}{8}\right)^2 + \dfrac{1169}{4} = 0$
Begin to isolate the variable.	$16(t - 2.875)^2 = 292.25$	$16\left(t - \dfrac{23}{8}\right)^2 = \dfrac{1169}{4}$

Divide both sides by 16.	$(t - 2.875)^2 = 18.265625$	$\left(t - \dfrac{23}{8}\right)^2 = \dfrac{1169}{4}$
Take the (positive) square root of both sides.	$t - 2.875 = \sqrt{18.265625}$	$t - \dfrac{23}{8} = \sqrt{\dfrac{1169}{64}}$
Add to both sides to get t.	$t = \sqrt{18.265625} + 2.875$	$t = \dfrac{\sqrt{1169}}{8} + \dfrac{23}{8}$

Either route leads to a solution of approximately 7.15 seconds, which students should identify as the time it takes for the rocket to hit the ground.

You can use this as an occasion to look back at the ideas from *Simply Square Roots* in *Do Bees Build It Best?*, asking students how they might simplify $\sqrt{\dfrac{1169}{64}}$. The purpose is not to get a "better" answer (the expression $\dfrac{\sqrt{1169}}{8}$ is really not any better), but to review some general principles about square roots.

Vertex Form for the General Quadratic

If it seems worthwhile and appropriate, you might have students look at putting the general quadratic function, $y = ax^2 + bx + c$, into vertex form, using their work on the height function from *Fireworks in the Sky* as a model. Even if you have to lead them through the process, they will see that the steps can be done for any quadratic function.

The steps might go like this:

$$y = ax^2 + bx + c$$

$$= c + a\left[x^2 + \left(\frac{b}{a}\right)x\right]$$

$$= c + a\left[x^2 + \left(\frac{b}{a}\right)x + \left(\frac{b}{2a}\right)^2 - \left(\frac{b}{2a}\right)^2\right]$$

$$= c + a\left[\left(x + \frac{b}{2a}\right)^2 - \left(\frac{b}{2a}\right)^2\right]$$

$$= c - a\left(\frac{b}{2a}\right)^2 + a\left(x + \frac{b}{2a}\right)^2$$

$$= \frac{4ac - b^2}{4a} + a\left(x + \frac{b}{2a}\right)^2$$

(The combining of terms in the last step is not essential to the big picture.) Students can then identify the vertex itself from the vertex form. (Setting the final expression equal to zero and solving for x is all that is left to derive the quadratic formula.)

Supplemental Activity

The Quadratic Formula (extension) asks students to apply and then derive the quadratic formula. You may want to go through the derivation with them.

A Fireworks Summary

Intent

Students summarize their work on the central unit problem, presented in *Victory Celebration*.

Mathematics

Students articulate the methods they used to address the unit problem.

Progression

This summative activity, to be done by students working individually, will produce a summary of the unit problem's solutions and solution methods.

Approximate Time

5 minutes for introduction

25 minutes for activity (at home)

Classroom Organization

Individuals

Doing the Activity

Tell students that this activity will be part of their unit portfolios. As with other summative activities, this is a chance for students to review what they have been doing for several weeks, to organize it in their minds and on paper, and to clarify any questions they might still have.

Intercepts and Factoring

Intent

These activities introduce another symbolic method for solving quadratic equations and ask students to summarize their work over the entire unit.

Mathematics

Factoring polynomials plays a large role in classical mathematics, persists as a major topic in traditional mathematics curricula, and appears on some standardized tests. Factoring is included in this unit so students can become familiar with the concept. It has been proven mathematically that almost all polynomials are prime and hence do not factor. For solving equations in applied situations outside of school mathematics, factoring is seldom used, and computers quickly estimate the roots of polynomials as accurately as needed. However, some polynomials can be factored, and when they can, it is easy to find the x-intercepts.

In *Intercepts and Factoring,* students use the area model for multiplication to factor quadratic expressions with leading coefficient of 1 or −1. They combine factoring with the **zero product rule** ($ab = 0$ if and only if $a = 0$ or $b = 0$) to solve quadratic equations and connect the solutions to the x-intercepts of the quadratic function. If a quadratic equation can be factored as a product of two linear terms, then it is easy to solve because of the zero product rule.

Progression

The activities support the development of factoring as a method for solving quadratics and then turn to completion of the unit portfolio.

Factoring

Let's Factor!

Solve That Quadratic!

Quadratic Choices

A Quadratic Summary

Fireworks Portfolio

Factoring

Intent

This activity introduces the technique of solving quadratic equations by factoring.

Mathematics

Factoring, when it is possible, is one way to find the solutions of quadratic equations. Factoring reduces a quadratic expression to the product of two linear expressions. This makes solving the equation easier because of the **zero product rule:** $ab = 0$ if and only if $a = 0$ or $b = 0$. If $ax^2 + bx + c = 0$ can be rewritten, in factored form, as

$$(px + q)(rx + s) = 0$$

then $px + q = 0$ or $rx + s = 0$. These linear equations are easy to solve. For simplicity, the work in this unit on factoring focuses on quadratic expressions in which the coefficient of x^2 is 1.

Progression

After a whole-class discussion of factoring, students work on this activity in groups.

Approximate Time

25 minutes

Classroom Organization

Whole class, then groups

Doing the Activity

Students may be familiar with the term *factoring* from work with whole numbers, such as writing 20 as 4 • 5. Many will recall drawing "factor trees" in middle school. They know, from working on the POW *Twin Primes,* that if a whole number has as factors only itself and 1, then it is prime. Numbers that are not prime can be written as a product of two or more factors greater than 1.

Explain that factoring in algebra has a meaning similar to factoring with whole numbers, but in algebra factoring applies to polynomials with whole-number coefficients. A polynomial is factored when it is written as a product of polynomials of equal or lesser degree. For example, students learned in *Distributing the Area II* that $x^2 + 3x - 10$ can be factored as $(x + 5)(x - 2)$ and in *Square It!* that $x^2 + 14x + 49$ is equivalent to $(x + 7)^2$.

Unlike multiplying, there is often no direct approach to factoring. Factoring polynomials usually involves making an educated guess about the factors and then checking the answer by multiplying the factors to see if they give the original polynomial.

Drawing an area diagram and entering what is known can help one to make an educated guess.

If students have trouble getting started, you might work through Question 1 together. To factor $x^2 + 6x + 8$, start with an area diagram like that shown in the student book and fill in the upper-left and lower-right rectangles. The upper-right and lower-left rectangles must then have a total area of $6x$.

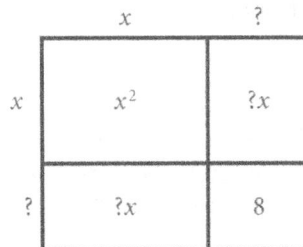

With some guess-and-check, students should see that if $x + 2$ and $x + 4$ are chosen as the dimensions, they will get the desired outcome. They can thus factor $x^2 + 6x + 8$ as $(x + 2)(x + 4)$.

Once we know the factors, it is easy to find the x-intercepts—the values of x for which the product is zero—by applying the zero product rule: a set of factors multiply to give zero if and only if one or more of the factors equals zero.

For $(x + 2)(x + 4)$, the first factor is 0 when $x = -2$. Then $(-2 + 2)(-2 + 4) = 0(2) = 0$. The second factor is 0 when $x = -4$. Then $(-4 + 2)(-4 + 4) = -2(0) = 0$. Therefore, the values of x that make the original quadratic $x^2 + 6x + 8$ equal 0 are -2 and -4. Of course, these values can also be found by completing the square.

Discussing and Debriefing the Activity

Allow time for several presentations that focus on the thinking that went into each solution.

You may want to give students another example in which the constant term is negative so that the x term must be thought of as a sum of a positive and a negative portion. For example, for the expression $x^2 - x - 6$, students will have to think of $-x$ as the sum $-3x + 2x$, leading to a diagram like this.

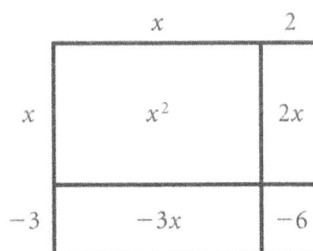

Also offer an example such as $x^2 + 3x + 1$, for which it is fairly clear that the expression cannot be factored, at least not using integer coefficients.

At the conclusion of this discussion, ask students whether there are any patterns or regularities that suggest a method for factoring quadratics.

Let's Factor!

Intent

This activity is a continuation of *Factoring.*

Mathematics

Students continue to develop their understanding of the factoring process as they use factoring to locate the *x*-intercepts of quadratic functions—the values of *x* for which a quadratic function has a *y*-value of zero.

Progression

Students work on the activity individually and share results with the class.

Approximate Time

20 minutes for activity (at home or in class)

15 minutes for discussion

Classroom Organization

Individuals, followed by whole-class discussion

Doing the Activity

No introduction is needed for this activity.

Discussing and Debriefing the Activity

For the examples that can be factored (Questions 1a, 1b, 1d, and 1e), be sure students see how to find the *x*-intercepts (Questions 2a, 2b, 2d, and 2e) from the factored expressions. For Question 1e, you might bring out that the two *x* terms in the related area model must total 0. Emphasize that if a quadratic expression can be factored, the graph of its related quadratic equation must have *x*-intercepts.

For Questions 1c and 1f, ask students to use vertex form or graphs to explain why the expressions can't be factored. Bring out these observations.

- The vertex form for Question 1c is $(x + 3)^2 + 1$ so the graph has its vertex at $(-3, 1)$ and is concave up. It therefore has no *x*-intercepts and thus can't be factored.

- The vertex form for Question 1f is $(x - 5)^2 - 19$ so the graph does have intercepts, but they are irrational, so factoring is not useful. The intercepts can be found by setting $(x - 5)^2 - 19$ equal to 0 and solving to get $x = 5 + \sqrt{19}$ and $x = 5 - \sqrt{19}$.

This unit is not intended to provide students with a full exposure to factoring techniques. The goal is for them to become aware of factoring as one possible method for solving some quadratic equations. They should realize that this method is very convenient if it works, but that it works only when the intercepts are rational (not square roots). If the intercepts are irrational, one can always use the vertex form approach.

Supplemental Activity

Let's Factor Some More! (reinforcement) encourages students to use an area model to factor quadratics in which the coefficient of the x^2 term isn't 1.

Solve That Quadratic!

Intent

This activity follows up on the second question of *Let's Factor!*, in which students connect solutions of quadratic equations to *x*-intercepts of the related graphs.

Mathematics

The values that make a quadratic equation equal to zero are the same as the values of the *x*-intercepts of the related graph. In this activity, students synthesize algebraic and geometric approaches to solving quadratic equations.

Progression

Students will work on this activity in groups to find solutions to quadratic equations and then to solve a quadratic equation drawn from a familiar context.

Approximate Time

5 minutes for introduction

20 minutes for activity (at home or in class)

10 minutes for discussion

Classroom Organization

Individuals or groups, followed by whole-class discussion

Doing the Activity

You might introduce this activity by reviewing the connection between solving quadratic equations and finding *x*-intercepts. Bring out that just as the number of *x*-intercepts for the graph of a quadratic function can be 0, 1, or 2, the number of solutions to a quadratic equation can be 0, 1, or 2.

Discussing and Debriefing the Activity

Students might approach the equations in Question 1 in various ways, but be sure the use of factoring is addressed. Questions 1a and 1b provide simple illustrations of that approach.

You can use Question 1c to illustrate the technique of simplifying an equation by dividing both sides by a constant. You might also bring out that although the equations $y = 2x^2 - 8x + 6$ and $y = x^2 - 4x + 3$ have different graphs, they have the same *x*-intercepts.

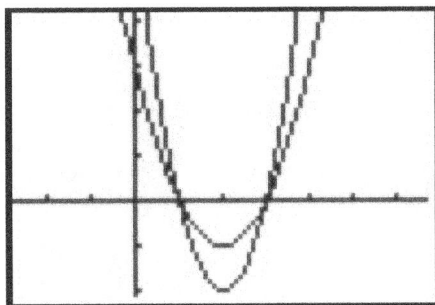

For Question 1d, be sure students explain why the equation has no solutions. They might do this using the graph or using the vertex form.

For Question 2, if students use the same labeling as in *Pens and Corrals in Vertex Form,* they should get the equation $x(500 - 2x) = 20,000$ for Question 2a. They might solve this in various ways, such as using a graph, vertex form, or factoring. Students should see several perspectives, such as that the line $y = 20,000$ crosses the graph of $y = x(500 - 2x)$ in two places.

Ask, **What do the two solutions represent in terms of the pen?** In this case, the two solutions represent different shapes for the pen. The solution $x = 50$ gives a 50-by-400 pen and the solution $x = 200$ gives a 200-by-100 pen.

You might lead a general discussion here about how students might decide what method to use for solving a quadratic equation. Among the considerations are these: Do they need an exact answer? How difficult is the algebra for the different methods? For instance, if an approximate answer will suffice, graphing may be the simplest approach. If students are doing a problem using an algebraic method, they might consider the factoring method if the expression is already in factored form or if the numbers seem easy to work with. On the other hand, as the vertex form method will always work, they might simply rely on this as their standard approach.

Supplemental Activities

Vertex Form Challenge (extension) gives students practice changing quadratic functions in standard form with leading coefficients other than 1 or −1 into vertex form. The problems involve fractions and decimals.

A Big Enough Corral (extension) explores quadratic inequalities. To do Question 2, students must be able to factor quadratics.

Factors of Research (extension) suggest further areas of exploration in the topic of factoring. Question 2 asks for a generalization of the difference of squares introduced in the supplemental activity *A Lot of Symmetry*.

Quadratic Choices

Intent

Students step back from the details of the two methods they have been learning for finding the x-intercepts of a quadratic function to reflect on when they might choose one method over the other.

Mathematics

Students now have three methods from which they can choose to find the x-intercepts of a quadratic function:

- Graphing, which will show how many intercepts and give their approximate values

- Vertex form, which will always give exact values for the x-intercepts, if they exist

- Factoring, which in certain limited cases can provide a quick way to find exact values for x-intercepts

The focus of this activity is on when to use each method.

Progression

Students will work on this activity individually. In Question 4, which will be the focus of the follow-up class discussion, students make general observations about the values of a, b, and c and how they might affect the choice of methods.

Approximate Time

25 minutes for activity (at home or in class)

25 minutes for discussion

Classroom Organization

Individuals, followed by whole-class discussion

Doing the Activity

This activity requires little or no introduction.

Discussing and Debriefing the Activity

Students will presumably want to use vertex form for Question 1 and factored form for Question 2, as the functions are in the given form. But there may be an interesting discussion for Question 3 and for the broader Question 4.

There are no right answers, and students may have legitimate disagreements about what's easiest. Be sure they see that the vertex form method will always work, even when factoring may be simpler.

Generalizing About x-Intercepts

For the general vertex form $y = a(x - h)^2 + k$, the number of x-intercepts depends only on a and k. This leads to these three general principles.

- If $k = 0$, there is exactly one x-intercept (no matter what a is).

- If k and a have the same sign, there are no x-intercepts.

- If k and a have opposite signs, there are two x-intercepts.

The teacher commentary for *Coming Down* notes that when the general quadratic function, written in standard form, is $y = ax^2 + bx + c$, then in vertex form it is

$$y = a\left(x + \frac{b}{2a}\right)^2 + \frac{4ac - b^2}{4a}$$

Students can apply the three general points above to this situation, where k is $\frac{4ac - b^2}{4a}$. You might bring out that the denominator $4a$ has the same sign as a, so the existence of x-intercepts depends only on the sign of $4ac - b^2$. If $4ac - b^2$ is 0, there is one x-intercept; if $4ac - b^2$ is positive, there are no x-intercepts; and if $4ac - b^2$ is negative, there are two x-intercepts.

Traditional presentations state these conclusions like this:

- If $b^2 - 4ac = 0$, there is one x-intercept.

- If $b^2 - 4ac > 0$, there are two x-intercepts.

- If $b^2 - 4ac < 0$, there are no x-intercepts.

Supplemental Activities

Make Your Own Intercepts (extension) builds on the idea that students can now easily find a quadratic that has two given x-intercepts. For example, for intercepts $x = 4$ and $x = 2$, the quadratic $y = (x - 4)(x - 2)$ will do. However, it isn't the *only* quadratic with those intercepts. All quadratics $y = a(x - 4)(x - 2)$ for any real number a will also work.

Quadratic Challenges (reinforcement) offers three more quadratic equations for students to solve. (A graphing calculator would make finding the requested decimal answers too easy.)

Standard Form, Factored Form, Vertex Form (reinforcement) pulls together the relationships among standard form, factored form, x-intercepts, vertex, and vertex form. This activity makes a good group assignment.

A Quadratic Summary

Intent

Students begin their portfolio preparation by summarizing the big ideas of the unit.

Mathematics

Some of the ideas students may discuss in their summaries are listed here.

- The graphs of quadratic functions, which are parabolas with symmetry and a turning point

- The connections between symbolic and graphical representations of quadratic functions

- Several methods for manipulating symbols to find key features of a function's graph, including the vertex and x-intercepts

- Methods for solving quadratic equations and their relationship to graphs of the corresponding functions

Progression

After a brief introduction, each student will work on his or her own to write a summary of the unit's big ideas.

Approximate Time

5 minutes for introduction

25 minutes (at home)

15 minutes for discussion

Classroom Organization

Individuals, followed by whole-class discussion

Doing the Activity

Tell students that this summary of the unit's big ideas will be part of their unit portfolios. To help them start thinking about the summary, ask volunteers to share ideas about some of the key concepts from the unit. Also mention the reference page in the student book, *A Summary of Quadratics and Other Polynomials,* which students may find helpful.

Discussing and Debriefing the Activity

As one aspect of the summary, ask students to discuss what sorts of situations they have seen in which quadratic functions occurred. Here are the main contexts from this unit; students may mention other possibilities as well.

- Gravitational fall, as in the unit problem, *Victory Celebration*

- Area (or volume with one dimension fixed), as in *A Corral Variation*

- Problems involving right triangles and the Pythagorean theorem, as in *Revisiting Leslie's Flowers*

- Problems involving maximizing profit, using the assumption that sales quantity is a linear function of price, as in *Profiting from Widgets*

Fireworks Portfolio

Intent

Students compile their unit portfolios and write their cover letters.

Mathematics

Portfolios for this unit include students' reflective work on the two activities *A Fireworks Summary* (in which they looked back over their work on the unit problem) and *A Quadratic Summary* (in which they reviewed the unit's big mathematical ideas). It also includes activities that helped them understand the value of vertex form in solving real-world problems and activities that helped them become comfortable with the mechanics of working with quadratic expressions.

Progression

Students start work on their portfolios in class by reading the instructions in the student book. They then work independently to review their work in the unit, select samples, reflect on the evidence of their learning, and write cover letters.

Approximate Time

10 minutes for introduction

35 minutes for activity (at home or in class)

Classroom Organization

Individuals

Doing the Activity

Have students read the instructions in the student book carefully.

By this point in the curriculum, students are familiar with the process of assembling a unit portfolio. Stress that this is their chance to identify any difficulties or questions they might still have about the mathematics of the unit. Remind students that their cover letters are important components of their portfolios, as they communicate with you what they have learned in the unit.

Discussing and Debriefing the Activity

You may want to have students share their portfolios in their groups, comparing what they wrote about in their cover letters and the activities they selected.

Assessments

In-Class Assessment

1. Write an equation for a quadratic function whose graph has its vertex at the point (−5, 3).

2. Write the expression $x^2 + 8x + 11$ in vertex form. Show how to use your work to get the vertex of the graph of the function $f(x) = x^2 + 8x + 11$.

3. What needs to be added to $x^2 + 18x$ to make it a perfect square? Explain your answer using an area diagram.

4. Write the quadratic expression $(x - 3)^2 + 8$ in standard form.

Take-Home Assessment

1. The zero product rule says that if a product of two numbers is zero, then at least one of those numbers must equal zero. Explain how the zero product rule is useful in solving equations.

2. Each graph here is the graph of a quadratic function. No scales are given, so all you know about a particular point is which quadrant it is in. The scales on the four graphs might not be the same.

a.

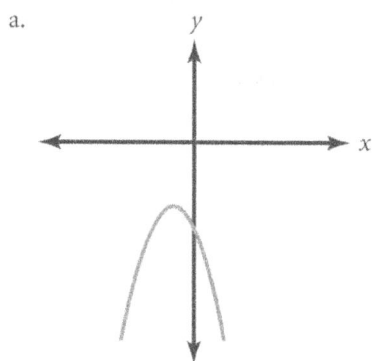

b.

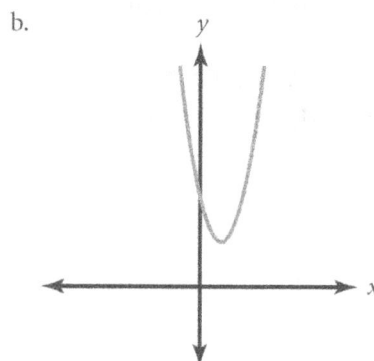

c.

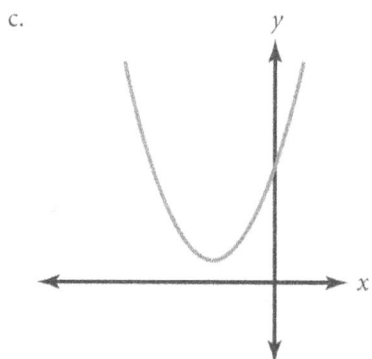

d.

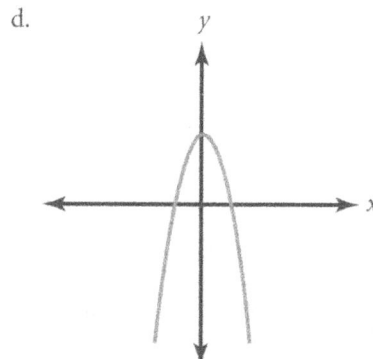

Match each graph with one of the equations listed here. For each graph, explain why that equation is the only one from the list that is appropriate.

i. $y = 3x^2 + 5$
ii. $y = 2(x - 1)^2 + 2$
iii. $y = -(x + 1)^2 - 3$

iv. $y = -x^2 + 6$
v. $y = -3(x - 4)^2 + 2$
vi. $y = \frac{1}{2}(x + 3)^2 + 1$

In Acapulco, Mexico, there is famous place called La Quebrada, or "The Break in the Rocks," where divers jump from high on the cliffs into the Pacific Ocean. Once a diver jumps, there is a relationship between how far from the cliff the diver is and how high above the water the diver is.

The relationship is given by the function $y = -(x - 0.5)^2 + 27.25$. In this function, x is the diver's distance from the cliff and y is the diver's height above the water. Both distances are in meters.

3. How high is the diver above the water when the diver is 3 meters from the cliff?

4. How far from the cliff is the diver when the diver is at the highest point above the water? How high is the diver at that time?

5. How far from the cliff is the diver when the diver is 7 meters above the water?

Fireworks Guide for the TI-83/84 Family of Calculators

This Calculator Guide gives suggestions for calculator use with selected activities of the Year 2 unit *Fireworks.* The associated Calculator Notes contain detailed calculator keystroke instructions that you can distribute to your students. NOTE: If your students have the TI-Nspire handheld, they can attach the TI-84 Plus keypad (from Texas Instruments) and use the calculator notes for the TI-83/84.

The *Fireworks* unit provides many opportunities for students to explore functional relationships using the graphing calculator, and to recall and practice many of the skills needed to work with this tool. Students will graph and create tables to investigate quadratic functions. Students can learn other features, like finding an *x*-intercept or a root, as they work through the curriculum.

Although much of the mathematics in this unit is traditional in nature, the opportunities for students to develop their understanding are enhanced by the use of the graphing calculator. Strong connections are made between symbolic and graphical representations of quadratic situations. Furthermore, students contemplate and assess the importance and distinction of approximate versus exact results.

Parabolas and Equations I: In the sequence of lessons *Parabolas and Equations I, II,* and *III,* students use graphing calculators to explore what happens to the graph of a parabola when the parameters *a, k,* and *h* in $y = ax^2$, $y = ax^2 + k$, and $y = (x - h)^2$ are changed. In this first activity, students explore the graphs of quadratic equations, and investigate how changes in the coefficients of a quadratic function affect its graph—that is, how the graph of $y = ax^2$ changes when the parameter *a* is changed. See the discussion in the *Teacher's Guide* activity notes for this activity. Additionally, this activity is an opportunity to refresh skills with graphing on the calculator. If your students need support with this, distribute the Calculator Note "Function Graphing".

Parabolas and Equations II: In this activity, students use the graphing calculator to explore how the graph of $y = ax^2 + k$ changes when the parameter *k* is changed.

Parabolas and Equations III: In this third activity, students use the graphing calculator to explore how the graph of $y = (x - h)^2$ changes when the parameter *h* is changed. Students have now done sufficient exploration to predict the appearance of any graph of an equation in the form $y = a(x - h)^2 + k$. They can quickly verify their predictions using the graphing calculator.

Vertex Form for Parabolas: In this activity, students use a graphing calculator to find the equation that produces a given graph. In this way, students gain facility with predicting both a graph based on an equation and an equation based on a graph.

Using Vertex Form: Students use the graphing calculator to further explore the relationships between graph and equation, and identifying the vertex. In Question 3, students estimate the coordinates of a vertex using a calculator graph. In doing this, it will be useful to adjust the window. This can be done quickly using the Zoom feature, and students will also use the Trace feature—both these features are discussed in the Calculator Note "Function Graphing."

Is It a Homer? In this activity, students will benefit from verifying their solution by graphing.

Square It!: In this activity, students change quadratic equations from vertex form to standard form. They can check their answers by graphing the two equations and verifying that they are equivalent—they should have the same graph. In Question 3, students can verify their answers by graphing.

Squares and Expansions: Students can verify their answers to Question s 2 and 3 by graphing.

POW: Twin Primes: If your students have done any calculator programming, you may want to suggest they write a program that gives all twin primes between two input numbers.

Squares and Expansions: Students can verify their answers by graphing.

How Much Can They Drink?: Students might use the graphing calculator to find the vertex and x-intercepts by tracing the graph.

Finding Vertices and Intercepts: Be sure that students practice finding vertices and intercepts using symbolic methods. The graphing calculator is useful for checking answers, but you may want to instead reinforce algebraic ways of verifying solutions at this point.

Another Rocket: To verify answers, you might have students graph the function on their calculators and then use the Zoom feature to estimate the x-intercepts. If their answers are correct, their answers should agree with the calculator values.

Coming Down: In this activity, students solve an equation to find when a firework hits the ground. You might wish to introduce students to estimating intercepts using a graphing calculator, using the Calculator Note "Determining x-Intercepts." You can have students check their work in future activities using these processes.

Solve That Quadratic!: This activity is a good opportunity to use graphs to reinforce the connection between solving quadratic equations and finding x-intercepts. Point out that just as the number of x-intercepts for the graph of a quadratic function can be 0, 1, or 2, so also the number of solutions for a quadratic equation can be 0, 1, or 2. Graph some of the equations and elicit that the solution to the quadratic equation are the values of the x-intercept.

Quadratic Choices: A third choice for finding the x-intercepts of a quadratic function is by graphing, which will show how many x-intercepts there are and will give approximate values for them. You may need to reiterate that you expect students to be able to find *x*-intercepts using symbolic methods, although graphing can be useful for verification.

Supplemental Problems:

Quadratic Challenges: There are three more quadratic equations to solve. The use of a graphing calculator makes finding the decimal answers too easy.

Calculator Notes

Function Graphing

These instructions describe the five basic graphing keys. You will find these keys immediately below the calculator's screen.

Using the $\boxed{Y=}$ Editor

Press $\boxed{Y=}$ to display this screen and enter functions. Use the $\boxed{X,T,\Theta,n}$ key for the independent variable *x* when you enter a function.

To remove a function from this list, move the cursor to that line (to the right of the = sign) and press $\boxed{CLEAR}$.

You can also make a function inactive without removing it. To do this, move the cursor over the = sign for that line and press $\boxed{ENTER}$. The = sign will no longer be highlighted. In the example shown here, the function **Y1=X²+5X+3** is active and the function **Y2=X²+5X+6** is inactive. The calculator will graph only active functions. To make the function active again, move the cursor to the = sign and press $\boxed{ENTER}$ again.

```
Plot1 Plot2 Plot3
\Y1◼X²+5X+3
\Y2=X²+5X+6
\Y3=
\Y4=
\Y5=
\Y6=
\Y7=
```

Setting the Viewing Window

Press $\boxed{WINDOW}$. This is where you tell the calculator what part of the graph to show and how to scale the axes. The **Xmin** and **Xmax** values determine the left and right bounds of the graph. The **Ymin** and **Ymax** values determine the bottom and top set at 1.bounds.The **Xscl** and **Yscl** values determine the frequency of the tick marks on each axis. **Xres** should be

```
WINDOW
Xmin=-8
Xmax=4
Xscl=1
Ymin=-5
Ymax=10
Yscl=2
Xres=1
```

Displaying a Graph

Press $\boxed{GRAPH}$. This tells the calculator to draw the graphs of the active functions in the $\boxed{Y=}$ edit screen. The portion of the graph you see will match the settings in the $\boxed{WINDOW}$ display.

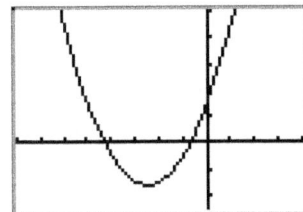

Exploring Graphs with ZOOM

Press ZOOM. The Zoom features are other ways to tell the calculator what part of the graph to show, and they automatically adjust the window settings.

6:ZStandard resets the window to a standard setting—*x* and *y* ranging from –10 to 10, with tick marks every 1 unit. To use **2:Zoom In** or **3:Zoom Out**, highlight your choice and press ENTER. You will return to the graph view, and a blinking cursor will appear at the center of the graph. Move the cursor to the spot you want to zoom in to (or out from) and press ENTER again. In the example below, **2:Zoom In** has been selected, and the cursor has been moved to near the parabola's vertex. Pressing ENTER zooms in on the vertex. Note that the coordinates of the cursor, shown at the bottom of the screen, remain the same.

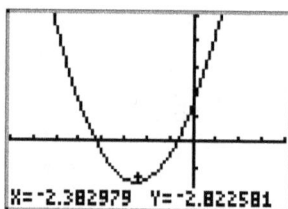

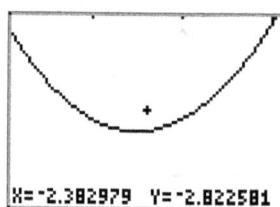

ZSquare adjusts your window to make the *x* and *y* units on the screen equally-sized, to keep your graphs from being stretched or distorted.

Exploring Graphs with TRACE

Once you have a graph displayed, press TRACE and press the left and right arrow keys to move the cursor along the graph. The *x*- and *y*-coordinates of the cursor are shown at the bottom of the screen. If you have graphed more than one function, use the up and down arrow keys to move the cursor between the different functions. The equation of the function the cursor is on is shown at the top of the screen.

You can use **Zoom In** in conjunction with TRACE to find the coordinates of a point on the graph, such as a vertex, with greater accuracy.

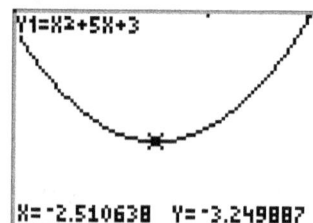

Determining x-Intercepts

You can use the calculator to solve equations when one side of the equation is zero. The examples here show several methods for finding (at least approximately) an x-intercept of the function $h(t) = 160 + 92t - 16t^2$ from the activity *Coming Down*.

To use the SOLVER feature, press MATH and then scroll down and select **0:Solver...**. The first screen should say EQUATION SOLVER across the top. If not, press the up arrow key once. Below the words EQUATION SOLVER, you'll find **eqn:0=**. Enter your function here. After typing in **160 + 92X – 16X²**, press ENTER.

```
EQUATION SOLVER
eqn:0=160+92X-16
X²■
```

```
160+92X-16X²=0
X=■
bound={-1ε99,1…
```

Your cursor should be at the **X=** as shown on the screen. Enter an estimate of where an x-intercept might be and then press ALPHA [SOLVE]. The next screens illustrate an example of what might result. Investigate what happens when you adjust the estimate you entered.

```
160+92X-16X²=0
X=7■
bound={-1ε99,1…
```

```
160+92X-16X²=0
•X=7.1488302493…
 bound={-1ε99,1…
•left-rt=0
```

A second method of finding an x-intercept is to instruct the calculator to determine the x-intercept while working from a graph of the function. To do this, press 2ND [CALC] and select **2:zero**. Next, the calculator will request guidance for its calculation algorithm by asking first for a left and right bound and then for a guess between the two. Move the cursor along the graph using the arrow keys, and press ENTER in response to each prompt. After these three prompts are entered, the calculator will display the x-value of the intercept.

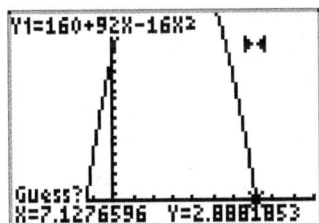

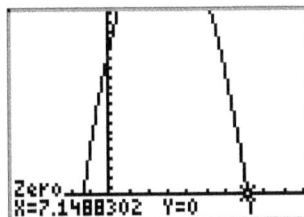

```
Y1=160+92X-16X²


Guess?
X=7.1276596  Y=2.8881853
```

```


Zero
X=7.1488302  Y=0
```

$[-5, 10, 1, -20, 200, 10]$

www.ingramcontent.com/pod-product-compliance
Lightning Source LLC
Chambersburg PA
CBHW051350200326
41521CB00014B/2528